DREDGING COASTAL PORTS

An Assessment of the Issues

Marine Board
Commission on Engineering and Technical Systems
National Research Council

NATIONAL ACADEMY PRESS
Washington, D.C. 1985

National Academy Press   2101 Constitution Ave., NW   Washington, DC 20418

This work is a result of research sponsored by the Water Resources Support Center, U.S. Army Corps of Engineers, under Contract Number DACW72-82-C-0010 between the U.S. Army Corps of Engineers and the National Academy of Sciences.

Library of Congress Catalog Card Number 85-62587

International Standard Book Number 0-309-03628-3

Printed in the United States of America

Marine Board

Contents

Preface

A subject that has long concerned the Marine Board of the National Research Council (NRC) is the systematic engineering of ports and harbors in the interests of efficiency, economy, navigational safety, and the protection of the marine and coastal environment. In reports addressing various aspects of this subject (National Research Council, 1981; 1983a; 1983b), the Board has noted the increasing importance and complexity of institutional issues. For the study of national dredging issues that is the subject of this report, the institutional issues proved particularly challenging. The study was conducted over a two-year period characterized by turbulent transition to a new federal policy for dredging—a transition that is as yet unresolved and difficult to predict—and severe distress in oceanborne shipping. The turmoil of this period produced a wealth of conflicting opinion and proposals for action; for the committee conducting the study, pressure to consider them all was intense.

ORIGINS OF THE STUDY

Acting on its charter "to consider questions of the relation of engineering and technology to navigation and the commerce of the sea and waterways," the Marine Board agreed in December 1981 to a request of the U.S. Army Corps of Engineers to appraise the nation's needs for dredging in the coastal ports for the near- and mid-term future.

STUDY SCOPE AND METHODS

A committee representing a broad spectrum of expertise was appointed to conduct the study under the direction of the Marine Board and to report its findings. Appendix A gives a brief summary of the members' expertise.

The committee was directed to investigate dredging needs in the coastal ports of the United States; that is, whether additional construction or maintenance dredging is needed now or in the next two decades, what impediments or barriers act against dredging, should such additional dredging be needed, and alternatives for responding to the impediments and barriers. The Marine Board asked that the committee consider in its appraisal: prospects for trade in various commodities and the vessels likely to carry that trade; alternatives to dredging; pertinent regulatory and institutional issues; design criteria for navigational channels, maneuvering areas, and emergency anchorages; the environmental effects of dredging and the disposal of dredged materials; and national security and defense. A selected bibliography covering these and related issues was prepared for the committee by the Study Center of the U.S. Maritime Administration's National Maritime Research Center.

Owing to the breadth of the issues, to the fact that all NRC committees are made up of volunteers, and to the deep divisions of opinion among experts and close observers in several of the areas being investigated, the committee employed a number of methods to meet its task. Principal among these methods were the review of evidence, preparation of background papers, and deliberations by the committee itself. Six meetings of the committee were held, four in conjunction with site visits and briefings in ports. An announcement of the study inviting comments was published in the *Federal Register*, April 20 and 21, 1983, and the responses were reviewed by the committee. The committee convened a public meeting in September 1983 to hear all interested views.

Several of the issues, in the committee's opinion, needed examination in depth.* Accordingly, the committee requested appointment of a technical panel through the NRC to examine the technical issues in design of dredged navigational facilities and the present adequacy of these facilities for the vessels and traffic using them. The work of the Technical Panel on Ports, Harbors, and Navigational Channels followed and amplified the work of an earlier panel (National Research Council, 1983b) that examined criteria for the depths of dredged navigational channels. The panel submitted its report to the committee following five meetings that were also coordinated with site visits and briefings. The panel's report was a principal source document for Chapter 6 of this report.

The committee also commissioned four papers from experts in various fields addressing (1) the biological effects of dredging and the disposal of dredged material; (2) the physical effects of dredging, control of sedimentation, and implications for sedimentation of new construction dredging, coastal structures, and natural events; (3) alternatives to dredging; and (4) consequences of various cost-sharing arrangements for major dredging projects. Information on national defense needs was requested from the Joint Chiefs of Staff.

While the committee conducting the study had much to consider and deliberate, the lack of data and analysis in oceanborne shipping must be noted. As the committee remarks in this report, considerable uncertainty attends the most fundamental question about national dredging needs—the question of demand for larger vessels. Some of the uncertainty can be attributed to the freedom and competitiveness of oceanborne shipping, some to the fact that political and economic decisions affecting this industry are often made far from its own sector, and some to the lack of regularly analyzed data (Office of Technology Assessment, 1983; National Research Council, 1984).

In contrast, considerable data have been collected and analyzed on the environmental effects of dredging and the disposal of dredged materials, and recent work has added to our understanding of vessel behavior in the confined waters of ports and harbors. Much remains to be learned in the latter area, however, and greater interdisciplinary communication is needed in all the areas examined by the committee to bring the results of research and development to bear on needed improvements in ports and vessels, and to the protection of the coastal environment. Collaboration is also needed to allow planning for the future.

In its own interdisciplinary effort, the committee represented strong opinions, and its deliberations were lively. Considering the great differences in experience and points of view the members brought to their common effort, and the turmoil of the two-year period during which the study was conducted, the committee achieved a remarkable level of agreement. This report represents the consensus of the committee.

*Background and commissioned papers are listed in Appendix F.

ACKNOWLEDGMENTS

The Marine Board gratefully acknowledges on behalf of the Committee on National Dredging Issues the generous contributions of time and information from the federal liaison representatives and their agencies, and the briefings and site visits provided by port authorities, local ship pilots, and other members of the maritime community. The background paper prepared by Betsy Gunn and the papers commissioned from Ray B. Krone, Ben C. Gerwick, Carl H. Oppenheimer, and James Sweeny and Edward Margolin proved particularly valuable to the committee's deliberations; these contributions are woven into the committee's report. Special thanks are extended to all those who communicated with the study by phone, by mail, and in person. The work of the National Research Council depends on the volunteer efforts of a wide community, and in the case of this study, the efforts volunteered were impressive.

REFERENCES

National Research Council, Marine Board (1981), *Problems and Opportunities in the Design of Entrances to Ports and Harbors* (Washington, D.C.: National Academy Press).

National Research Council, Marine Board (1983a), *Collisions of Ships and Bridges: The Nature of the Accidents, Their Prevention and Mitigation* (Washington, D.C.: National Academy Press).

National Research Council, Marine Board (1983b), *Criteria for the Depths of Dredged Navigational Channels* (Washington, D.C.: National Academy Press).

National Research Council, Committee on National Statistics (1984), *Statistics for Transportation, Communication, and Finance and Insurance: Data Availability and Needs*, Staff paper prepared for the Committee by S. D. Helfand, V. Natrella, and A. E. Pisarski (Washington, D.C.: National Academy Press).

Office of Technology Assessment, U.S. Congress (1983), *An Assessment of Maritime Trade and Technology* (Washington, D.C.: Government Printing Office).

DREDGING COASTAL PORTS

An Assessment of the Issues

1
Executive Summary

The question whether ports in the United States are adequate to serve
the nation's present and future needs became a major public concern in
the early 1980s. Attention focused on port adequacy when, as a result
of the Iranian Revolution, world demand for U.S. coal exploded.
During 1980, news media in the United States were full of reports that
large numbers of colliers were waiting for weeks and sometimes months
to gain access to U.S. coal-loading facilities. During this same
period, a number of studies concluded that the United States had the
opportunity to become a major supplier of a large new world market for
steam coal. To gain and secure that market, it was repeatedly argued,
the United States would need to be able to handle the most efficient
dry-bulk carriers, and such carriers require greater water depths than
those available at U.S. coal ports.

The events of the early 1980s brought to public attention an issue
that had long been developing. The issue had two components. First
was the growing interdependence of the U.S. and the world economies.
During the 1960s and 1970s, the U.S. economy moved from being
essentially self-contained to becoming the largest component of a
world economy. Increasingly, U.S. economic well-being was seen as
being dependent on the nation's capacity to compete in a world
economy. Particularly for high-volume, low-cost commodities such as
coal, efficient low-cost transportation was viewed as an essential
ingredient to American competitiveness. For such commodities, large
bulk carriers were believed to offer major economies of scale. Many
of those involved in ocean transportation noted that the United States
would only be able to enjoy these economies of scale if it developed
substantially deeper ports, ports capable of handling large deep-draft
vessels.

Second, although the seeming advantages and trends to larger
vessels in the world fleet were evident, the 1970s saw little in the
way of a response to these perceived needs for deeper U.S. ports. The
inability of the nation to respond to the apparent need for deeper
ports was the result of an unraveling of the social contract that had
been in place for over 150 years between the federal government and
the ports concerning how both maintenance and new construction
dredging would be funded, managed, and regulated. By the 1980s, then,

1

24 ports had proposals for improvement dredging projects and no significant new construction dredging had occurred for a decade.

This report is an investigation of the major issues associated with port dredging. Specifically, it investigates three general questions: (1) Is additional port construction and maintenance dredging necessary now or over the next two decades? (2) What impediments and barriers militate against carrying out additional dredging if it is needed? (3) What alternatives offer promise of mitigating or effectively responding to those impediments in order that any needed dredging can be carried out?

Assessing the nation's dredging needs requires setting them in a more general context. It is necessary to seek an understanding of the role of ports in the broader U.S. and world economy and ocean transportation system. Further, it requires comparing the dredging of existing ports with a variety of alternatives that have been proposed for meeting the nation's transportation needs. Proponents and opponents of port dredging and the alternatives to dredging range across a broad spectrum. Some contend that immediate dredging of existing ports is a necessity while others argue that U.S. ports are adequate for the foreseeable future or that there are more cost-effective ways than dredging to meet the nation's need to handle large vessels.

The central conclusion of this report is that the nation needs additional capacity to handle large vessels and that such a capacity should exist on each of the nation's coasts. It is important to emphasize two reasons for this conclusion. First, the United States faces great uncertainty with regard to the size and character of the future world economy, the nature of future oceanborne transportation into and out of U.S. ports, and the future mix of commodities that the nation will export and import. Further, the character of U.S. exports and imports, particularly exports of such bulk commodities as coal and agricultural products, is likely to fluctuate greatly from year to year. Most of the analyses and arguments used by proponents and opponents of additional port capacity to handle large ships start from assumptions about the future size and character of U.S. trade and transport. All these assumptions must be viewed as highly uncertain. Decisions with regard to developing additional port capacity, then, must be made with the recognition that future needs are difficult to determine.

There is less uncertainty about the time required to develop additional port capacity. Port construction requires long lead times that will be measured in years. In the case of major federal dredging projects, the lead time is now 22 years. The nation, then, faces a fundamental mismatch between the uncertain and fluctuating character of future need and the certain and long times required to develop additional port capacity. A decision to develop additional capacity, therefore, involves risks. Unless those risks are taken, however, the United States precludes the opportunity to take advantage of any benefits offered by large ships in the future. It must be emphasized that particularly with regard to bulk commodities, extreme swings in trade and transport have traditionally occurred over very short periods of time. To take advantage of rapidly expanding markets, the nation must have available port capacity when those swings occur. To protect against the loss of those markets when world demand declines, the nation must be able to offer its products at the lowest possible cost.

The major findings of the study immediately succeed this summary. It must be noted that the key finding--the nation needs additional port capacity--is not derived from a detailed economic analysis. Rather, the finding represents the committee's consensus judgment of what is in the nation's interest, given an uncertain future. The committee found only disagreement in its review of existing research and in its interviews with experts concerning future port needs and the economic benefits of deeper ports. Thus, there is no consensus in the expert community on the costs and benefits of deeper ports. The committee chose to frame the central question of its study, then, as "What should the nation do if future port needs are uncertain?" and concluded that in the face of uncertainty it is prudent to have increased options--that capability should exist to enjoy maximum benefits given a wide range of future developments.

A common reading of several developments led to this consensus judgment--the growing importance of world trade to the economic well-being of the United States; a trend to larger ships because they offer economies of scale; the importance of ocean transportation costs in the delivered price of high-volume, low-cost commodities, such as coal; the growing number of deep-water ports in other countries; rapid year-to-year fluctuation in trade in particular commodities; and the long lead times required to develop deep port capacity, and thus, the inability to develop additional port capacity in response to short-term fluctuation and need.

The report is organized into seven chapters. Chapter 1 provides an overview of the background and issues associated with port dredging. The following six chapters investigate the six basic issues which the committee found must be resolved if the port adequacy question is to be meaningfully addressed. Those six questions are as follows: (1) Does the United States need additional port capacity to handle large ships? (2) Is dredging the most attractive way for the United States to handle large ships? (3) How should dredging be funded and what are the implications for dredging of various funding approaches? (It should be emphasized that the committee, in defining this task, excluded overly specific funding recommendations. Resolution of the funding issue is inherently a political choice, which in the system of the United States must be made by Congress.) (4) What are the causes of the slowdown in decision making for local port projects and the stalemate for federal projects, and what are the ways to bring increased speed, predictability, and stability to the decision making process? (5) What are the problems associated with the design and implementation of new construction dredging and how can they be dealt with? (6) What are the environmental problems associated with dredging, and can they be effectively managed?

The 33 findings that follow are organized as responses to each of these questions.

2
Findings

1. DOES THE UNITED STATES NEED ADDITIONAL PORT CAPACITY TO ACCOMMODATE LARGER SHIPS?

Finding 1: The United States should act expeditiously to increase its capacity to handle large ships. Two general categories of need exist: first, the capacity to accommodate liquid- and dry-bulk carriers of between 90,000 DWT and 150,000 DWT (or more); and second, a capacity to accommodate such modern vessels as the latest-generation containerships, which require water depths of 40 ft to 45 ft. Several facts and trends lead to this finding:

- Large vessels offer economies of scale.
- Large liquid- and dry-bulk carriers (100,000 DWT and more) now dominate world trade in several commodities, and are rapidly increasing their share of other commodity movements.
- Large liquid- and dry-bulk carriers cannot transit the major ports of the Atlantic or Gulf coasts fully loaded. (There is some deep-water loading/unloading capability on the Pacific Coast.)
- Many major ports of the United States cannot easily accommodate the range of medium-size vessels--for example, those designed to the maximum dimensions allowed by the Panama Canal, and the latest-generation containerships--a range requiring water depths of 40 ft to 45 ft (or more).
- A substantial number of ports worldwide have the capacity to accommodate large ships, and several more are planning or developing this capacity.
- Some combatant vessels, and (increasingly) the commercial vessels that would be relied on to support any major defense mobilization require greater water depths or widths than are now provided in some designated defense ports.
- Future patterns of trade and the vessels that will carry that trade are uncertain. In the face of uncertainty, prudence dictates having sufficient flexibility to respond rapidly to a range of possible developments.

2. WHAT ARE THE MOST ATTRACTIVE WAYS FOR THE UNITED STATES TO ACCOMMODATE LARGE SHIPS?

Finding 2: The preferred options for developing additional capacity to handle large ships are: (1) dredging existing multicommodity ports and (2) lightering/topping-off.

   For a range of vessels and cargoes, there is no adequate alternative to conventional ports and harbors for loading or unloading. Existing ports offer sheltered water for unloading or loading without delay, access to developed storage and cargo transfer facilities, well-developed connections to inland transportation, and established worldwide marketing networks. Dredging to increase the capacity of existing ports yields maximum future flexibility to respond to changes in trade, and to support defense mobilization, as well as offering the broadest base for recovering the cost of dredging and distributing the benefits.

   Several alternatives to dredging existing ports are specific to one or more bulk commodity. The attractiveness of these options depends on several factors such as location, commodity, and volumes to be handled that can only be evaluated or compared in specific proposals.

   The lowest-cost alternative to dredging is lightering/topping-off. This alternative already exists in the United States, and offers a flexible response to the need to load and unload bulk carriers in some contexts.

Finding 3: The nation needs to assure, on an accelerated basis, the existence of permanent, multipurpose port capacity to handle bulk vessels requiring at least 50 ft of water depth on each of the coasts.

Finding 4: If federal funding is involved, the choice of which ports to deepen to 50 ft or 55 ft or more is particularly important because the high cost of port construction and maintenance and present uncertainty associated with need suggests that only a limited number of such port deepening activities are warranted. Selection of port deepening projects should consider four criteria: (1) Emphasis should be given to dredging ports that handle all types of cargoes, and therefore, a variety of vessels--containerships, roll-on/roll-off ships, break-bulk vessels, and both liquid- and dry-bulk carriers. (2) Major attention should be paid to the port's inland transportation systems. Two factors are important here. First, the most attractive ports would be those serving a broad range of economic activities, from manufacturing to agriculture to mining, as well as serving large populations or markets. Second, wherever possible, the port should be served by multiple inland transportation systems. The ideal would be ports served by multiple rail lines, highway transportation, and inland waterway transportation. Such multiple inland transportation options offer a competitive environment which should serve to keep inland transportation costs low. (3) Consideration should be given to the costs of construction and maintenance dredging. Particular emphasis should be placed on minimizing maintenance dredging since it is a recurring cost. (4) Ports should be judged in terms of the

environmental consequences of dredging. To the extent that
environmental consequences can be reduced and other criteria
satisfied, the optimum situation would exist.

Finding 5: The navigational facilities of some major ports need to be
enlarged by dredging to handle the range of medium-size vessels
requiring 40 ft to 45 ft of water depth (or more).

Dredging existing ports to handle Panamax vessels and the
latest-generation containerships requires less dredging than that
required for larger vessels. If federal funding is involved, emphasis
should be given to ports that handle all types of cargoes, as a
variety of vessels in the world fleet require water depths of 40 ft to
45 ft. Attention also needs to be given to the ports' inland
transportation system, to populations and inland markets served, to
the comparative costs among the candidate ports of construction and
subsequent maintenance dredging costs, and to the environmental
consequences (that is, emphasis needs to be given to the reduction of
potentially adverse consequences and the demonstration of
environmental benefits in selecting dredging projects).

Finding 6: Lightering/topping-off should be encouraged.
Even on an accelerated basis, dredging projects will take time.
Where needs now exist to accommodate large bulk carriers,
lightering/topping-off can be implemented with speed and flexibility.
Private-sector investments have already been made in these systems,
and where their use is attractive, it should be encouraged.
In sum, the committee concludes that the nation will have adequate
flexibility to handle bulk commodities in the future with (at a
minimum) deep-draft capacity on each coast and lightering/topping-off
capability, and medium-draft capacity in some major ports.

3. HOW SHOULD DREDGING BE FUNDED, AND WHAT ARE
THE IMPLICATIONS FOR DREDGING OF
VARIOUS FUNDING APPROACHES?

Finding 7: Some form of federal funding is warranted, given the
importance of enhanced port capacity to the nation's economic
well-being, and national security/defense needs.
Ports are a national resource. The U.S. and world economies are
increasingly interdependent, and most trade among nations is
oceanborne. Enlarging the physical capacity of U.S. ports could make
our exports more competitive and reduce the landed cost of imports.
These conclusions, as well as the potential role of ports in national
security and defense, argue for some continued federal funding.

Finding 8: Congressional consideration of port projects should be
independent of other water resources projects, and Congress should
identify the criteria to be used in selecting among competing projects
for which there is a federal funding role.

Finding 9: Resolving the funding stalemate will require addressing
the following issues: (1) the sources of revenue, (2) who will
collect the revenues, (3) how the revenues will be allocated, (4) who
will handle the management and implementation of port dredging, and
(5) what the preceding four changes will mean for the management and
regulatory approval process associated with port dredging projects.
It is important that Congress address all these issues if port
dredging is to be expedited.

4. WHAT ARE THE CAUSES OF THE SLOWDOWN IN DECISION MAKING
FOR LOCAL PROJECTS AND THE STALEMATE FOR FEDERAL PROJECTS,
AND WHAT ARE THE WAYS OF BRINGING INCREASED SPEED,
PREDICTABILITY, AND STABILITY TO THE DECISION MAKING PROCESS?

Finding 10: The complexity of the institutional decision making
system for port dredging reflects the need to address and manage a
large and difficult set of real needs and concerns. But the system
does not allow timely port development, either of federally or locally
funded projects. Actions need to be taken by Congress, the regulatory
agencies, and the ports to achieve speed, predictability, and
stability in decision making, as suggested in succeeding findings.

Finding 11: Ports need to establish mechanisms and procedures for
developing and evolving comprehensive port plans.
    In general, regulatory decisions governing port dredging have only
been made when consensus is achieved--that is, when no significant
participants have objected so strongly to proposed action that they
are willing to mobilize and oppose it. A planning process is the
essential beginning of continuing consensus-building for ports.

    Any planning process needs to identify the needs of the port both
in the short and long term and the implications of those needs for the
range of concerns reflected by the interested parties. The planning
mechanism or process, then, needs to include all the appropriate
governmental agencies as well as port users, commercial interests, and
environmental and public interests concerned about port development.
Since each port is unique, what is required will vary from port to
port, but given the complexities of the issues that now surround port
decision making, any route other than a local consensus building
mechanism appears to offer little chance of success.

Finding 12: Resolving environmental issues in a timely fashion has a
direct bearing on the economics of a dredging project. Delays
occasioned by the actual or perceived failure to address environmental
concerns adequately can result in significantly increased costs, which
may undermine or negate the economic benefits expected from the
dredging projects--and even in the abandonment of worthwhile projects.

Finding 13: In the interests of preserving the right of those with
strongly held objections to have their day in court, while not

subjecting project proponents to unreasonable and protracted opposition, Congress should consider placing appropriate time limits on the ability to seek judicial review once a final regulatory decision has been made on a port dredging project.

Finding 14: A review of existing dredging-related regulations needs to be undertaken to determine their consistency with the intent of federal law, their necessity, efficacy, and clarity.

Finding 15: New procedures are needed to allow the comments of state and federal agencies on proposed dredging projects to be consolidated to the extent possible, both for the review of environmental impact statements and for the review of permit applications (that is, for both federally and locally funded projects).

Finding 16: A comprehensive interagency agreement needs to be developed for consistent and timely consideration of environmental mitigation by all agencies commenting on dredging or filling projects under the Fish and Wildlife Coordination Act. In addition, each U.S. Fish and Wildlife Service region should be given the authority and responsibility to develop a regional mitigation plan with specific resource management goals to guide the formulation of project-specific mitigation requirements.

Finding 17: In the formulation of project-specific requirements, regulations should clearly state that wildlife resource agencies include and consider the economic costs and environmental benefits of mitigation proposals. At a minimum, these agencies need to demonstrate that mitigation (1) will achieve significant benefits for environmental resources within the statutory authority of the commenting agency, (2) is reasonably available to the applicant, and (3) is feasible and practicable.

5. WHAT ARE THE PROBLEMS ASSOCIATED WITH DESIGN AND IMPLEMENTATION OF DREDGING PROJECTS, AND HOW CAN THEY BE DEALT WITH?

Finding 18: The design of new construction and dredging projects is impeded by the length and character of the decision making process.
   As proposals age, the world fleet and port structures change, but engineering redesign is confined to the limits of project authorization, if the proposal is approved. Systematic port engineering is discouraged by the length and fragmentation of institutional processes.

Finding 19: Safety margins that would otherwise be provided by engineering are in many ports now provided by compensating operational practices.
   Because there are as yet no consensus standards for navigational safety, the design of new construction and ongoing maintenance

9

dredging projects should be carried out with full cooperation and coordination among all relevant interested parties: the U.S. Army Corps of Engineers, U.S. Coast Guard, harbor pilots, shipowners and other shipping interests, to ensure local consensus on navigational safety.

Finding 20: Design, layout, and dimensions of dredged navigational facilities must be site-specific. The consensus standards developed by international organizations should be adopted as general guidelines, where applicable. Better understanding can and should be gained of vessel maneuvering characteristics, and efforts need to be made to improve and use the techniques and tools of design analysis for accommodating vessels; e.g., the full-scale vessel simulator of the U.S. Maritime Administration. Techniques and tools are well developed for observing and modelling local circulation and sedimentation (cf. Finding 24). The use of these sets of tools needs to be encouraged.

Finding 21: The use of existing state-of-the-art dredging plant needs to be encouraged, and impetus given to developing needed modern capability.

Finding 22: Contracting with private dredging companies should be handled to achieve maximum efficiency and optimum use of resources.
   Improvements can be made by providing long-term dredging contracts that allow dredging contractors to make investments in equipment offering greater efficiency. Review of procurement and other procedures may suggest further revisions to reduce costs, maximize efficiency, or both.

6. WHAT ARE THE MAJOR ENVIRONMENTAL PROBLEMS
ASSOCIATED WITH DREDGING AND
HOW CAN THEY BE MOST EFFECTIVELY MANAGED?

Finding 23: Since dredging leads to disposal, the two activities are tightly coupled. The potentially undesirable environmental effects of the two activities are quite different, however.
   The potential for persistent undesirable effects associated with the dredging of materials for maintenance is very small, regardless of the character and quality of materials dredged. The potential for persistent undesirable effects associated with the disposal of these materials may be significant. It is a function of the quality of materials dredged, their physical characteristics, and the levels of associated contaminants.
   The potential for persistent undesirable effects associated with the dredging of material for new work may be significant. It is a function of the quantity of material dredged, the changes in channel geometry, and the local hydraulic regime. The potential for persistent adverse environmental effects associated with the disposal of this material, on the other hand, is small.

Scientific understanding of the environmental effects of dredging and dredged-material disposal is sufficient to enable predictions of the short- and long-term effects of maintenance and construction dredging, and, at least, the short-term effects of disposal with a level of accuracy sufficient to ensure adequate environmental protection.

Finding 24: The greatest potential for environmental problems associated with dredging is with construction dredging, which can produce persistent and significant changes in the hydraulic regime as a result of channel deepening. The potential for such effects requires particular attention in estuaries, where changes in cross-sectional geometry can alter estuarine circulation patterns and as a result, the distribution of salinity, dissolved oxygen, and other important environmental parameters.

The greatest potential for environmental problems associated with disposal of dredged materials is associated with maintenance materials which contain moderate to high concentrations of toxic material. Typically, this contaminated fraction constitutes a relatively small percentage of materials removed, and enough is known to select appropriate disposal options to contain the contaminated material and thereby minimize the potential for adverse effects.

Finding 25: Sedimentation within many port areas is significantly affected by sediments introduced by the activities of man, including construction, farming, and a variety of municipal and industrial discharges. Greater efforts should be made at the federal, state, and local levels to control these sources throughout the tributary drainage basin.

Finding 26: Parallel efforts should be made to reduce or eliminate inputs of toxic contaminants entering drainage basins and rivers or estuaries leading to ports. The costs of controlling toxic contaminants at the source may very well be less than those associated with the disposal and management of toxic materials dredged from ports. For existing contaminated sediments, a properly designed and executed program of dredging and containment may have beneficial environmental consequences.

Finding 27: A comprehensive dredging and dredged-materials management plan should be developed for each port with a specific long-term objective being to assure that maintenance projects can be carried out on schedule while minimizing adverse environmental effects.

The plan should be based on: (1) a thorough characterization of the kinds and qualities of material to be dredged, (2) a detailed determination of the spatial distribution of contaminants (both horizontally and vertically) within channel deposits that permits definition of the degree of homogeneity in the sediments to be dredged and to delineate prominent contaminant "hot spots," (3) a rigorous assessment of the physical and chemical behavior of these materials if placed in each of the alternative disposal environments, (4) consideration of beneficial use (cf. Finding 29) as an alternative, (5) an assessment of the effects resulting from each of the dredging and disposal alternatives for public health, the environment, the biota, other uses of that segment of the environment, and the relative costs, and (6) consideration of long-term continuing costs and effects associated with the plan.

Such a plan should eliminate the need for a complete new assessment every time a maintenance project is scheduled. Given that the material accumulating within any development area varies little over relatively long periods of time (unless there is a major spill or other accidental release of contaminants, or natural event such as a flood or violent storm), analysis of representative materials should be sufficient to determine whether the quality of sediment falls within the normal range. Careful development of a dredging and dredged-materials management plan should reduce, if not eliminate, the need for repeated bioassay and bioaccumulation tests.

Finding 28: From the point of view of environmental protection, a major problem associated with regulatory procedures and criteria is their lack of responsiveness to new information about the environmental effects (positive and negative) of dredging and the disposal of dredged material. Far more is known about environmental effects and probable causes than is incorporated in present regulatory criteria. Thus, streamlining the regulatory process is necessary--not just for port management, but for the improvement of regulatory criteria.

Finding 29: Dredged sediment should be regarded as a resource rather than as a waste.

Materials should be carefully screened to determine suitability for use as construction aggregate, sanitary landfill cover, for beach replenishment, for the creation and enhancement of wetlands, and for other uses prior to disposal. The attractiveness of these alternatives, as well as a balanced assessment of upland sites versus ocean sites for disposal, would be encouraged if the "local cooperation" policy of the U.S. Army Corps of Engineers were modified. That policy requires that the project proponents assume responsibility for "all necessary lands, easements, and rights of way" for upland sites in addition to ongoing operation and maintenance costs. Such responsibilities, not required of users of ocean disposal sites, limit the attractiveness of the upland alternative independent of benefits that might be realized.

Finding 30: Designated containment sites, whether specially constructed (e.g., diked alongshore structures, containment islands, or upland containment facilities) or remnants of previous commercial activity (e.g., subaqueous borrow-pits), should be reserved for the disposal of dredged materials known to be contaminated with toxic compounds.

Contaminant-free sediments should be placed within these areas only as cover or "capping" materials, and every effort should be made to minimize the volume occupied by these clean materials. Storage of contaminated sediments in these sites should be managed so as to maintain a physically and geochemically stable environment, to minimize the exposed surface area of the deposit, and to preserve essentially anaerobic conditions.

Finding 31: Characterization and designation of ocean disposal sites by the Environmental Protection Agency should be completed as rapidly as possible.

The present use of historical disposal areas does little to minimize the probability of adverse short-term effects and may ultimately result in long-term continuing problems. So long as ocean disposal sites are not officially designated, it is difficult to justify detailed analyses of sediments, since these analyses have little influence on management of the sediments.

Finding 32: The approach that characterizes many studies of the environmental effects of dredged-material disposal needs to be modified.

There should be reduced emphasis on survey techniques (counting, sorting, diversity, and the like) and more emphasis on biological response, including physiological processes. There is a particular need for some reasonably long-term studies modelled on the lines of ongoing public health or epidemiological research. Associated with the initiation of these studies should be the termination of the majority of the short-term investigations typically associated with Environmental Impact Statements (EIS). While the EIS process continues to represent a valuable component of over-all environmental management and regulation, it has been reasonably well demonstrated that routine field and laboratory studies lack the resolution required to detail all but the most evident acute effects.

Finding 33: Procedural specifications used within federally sponsored dredging projects should increasingly encourage the use of high-efficiency dredges and dredging techniques specifically intended to reduce sediment loss and associated turbidity during dredging and to provide routine, highly accurate placement of dredged materials.

Where predredging surveys indicate contaminated "hot spots," selective dredging and disposal techniques should be used. These techniques need to be considered in contracting for dredging plant (cf. Findings 21 and 27).

3
Overview of Dredging Issues

The adequacy of American ports to meet the nation's present and future needs became the focus of major policy attention in the early 1980s. The issue attained sufficient importance to be addressed both by President Reagan in his 1983 State of the Union message and in the Democratic party's response to that message (Democratic National Committee, 1983). Two complex developments converged to focus this high-level attention on the adequacy of ports. The first was the changing character of the U.S. economy vis a vis the world economy, and the changing character of commercial shipping into and out of the nation's ports. The second development was the unraveling of a social contract that had evolved over 150 years between the federal government and the ports concerning how both maintenance and new construction dredging would be funded, managed, and regulated. By the early 1980s these two general developments were characterized by:

- Proposals from 24 ports for improvement dredging projects based on the perception that such capabilities were critical to their future competitiveness.
- More than a decade of paralysis for federally funded new construction dredging and major delay for some ports willing to fund their own construction dredging.

The most powerful pressure for developing the capacity to handle large ships is the claim that such ships offer lower-cost transportation. Advocates of additional port dredging contend that without the capacity to handle large, economically efficient ships, commerce into and out of the United States must either use smaller, higher-cost ships or larger ships must enter and leave the nation's ports less than fully loaded. In either case, higher transportation costs are the result. These higher transportation costs, it is argued, have the effect of increasing the price of American products on the world market and raising the landed price of foreign imports.

The controversy over ports revolves around dredging. Resolution of that controversy requires finding answers to six issues: (1) Does the United States need additional port capacity to handle larger ships? (2) Is dredging the most attractive way for the United States to handle larger ships? (3) How should dredging be funded and what are

13

the implications for dredging of various funding approaches? (The committee in defining its task specifically excluded recommendations about funding formulas, since this is a political choice which will be made by Congress.) (4) What are the causes of the slowdown in decision making for local projects and the stalemate for federal projects, and what are the ways to bring increased speed, predictability, and stability to the decision making process? (5) What are the problems associated with design and implementation of new construction and maintenance dredging and how can they be dealt with? (6) What are the environmental problems associated with dredging and how can they be most effectively managed?

Traditionally, dredging has been divided between federal and local projects. Federal projects are paid for with congressionally appropriated funds and are carried out by the U.S. Army Corps of Engineers. In general, federal projects deal with the construction and maintenance of major access channels, maneuvering areas, and emergency anchorages in U.S. ports.

Local projects are characterized by nonfederal funding and generally deal with construction and maintenance dredging of berths and minor channels, or landfill projects (or both). Local projects do not require congressional action and are normally not managed by the Corps, but they are subject to detailed regulatory review which rests primarily with the Corps.

Construction dredging normally involves creating new navigational facilities or the improvement of those that exist by underwater excavation. Maintenance dredging involves the removal of materials as necessary to keep facilities at the originally constructed depths and widths. Although the physical activities required to carry out these two types of dredging are similar, the issues associated with them may be quite different. Differences range from how the decisions to dredge are made through how the dredging is funded, to regulatory approval procedures. Although controversy surrounds both maintenance and construction dredging, clearly construction dredging--specifically whether there is a need to handle larger, deeper-draft vessels, and if so, who should pay for it--is the key issue driving the present national debate.

The subject of this report is an investigation of several issues associated with port dredging. It is organized around three general questions: (1) Is additional port construction and maintenance dredging necessary now or over the next two decades? (2) What impediments and barriers militate against carrying out additional dredging if it is needed? (3) What alternatives offer promise of mitigating or effectively responding to those impediments in order that any needed dredging can be carried out?

Proponents of new or additional port dredging have identified three areas of need. The first is for additional capacity to handle deep-draft ships and more traffic. The deep-draft ships most commonly referred to are liquid- and dry-bulk carriers requiring water depths of 50 ft or more. The second is the need in a number of ports for additional depths to handle ships requiring water depths of 40 to 45 ft. (The vessels most often cited in this category are latest-generation containerships, but a number of vessels in the world fleet require water depths in this range. The third need is for additional maintenance dredging.

Assessing the nation's dredging needs requires setting them in a more general context. Specifically, it is necessary to understand the role of ports in the broader world economy and transportation system to assess whether additional capacity is needed, and if needed, whether there are alternative ways of meeting the nation's port requirements that are more attractive than dredging. Proponents and opponents of port dredging and the alternatives to dredging range across a broad spectrum. Some contend that U.S. ports are adequate; others contend that there are more cost-effective ways than dredging to meet the nation's need to handle large vessels.

In assessing present and future port needs and ways of meeting those needs, this report applies five broad criteria: economics, navigational safety, environmental implications, national security and defense, and implications for future ocean transportation flexibility.

Ports are one component of a five-component international or coastwise transportation system. In the case of exports, the first component is the inland transportation network that carries goods from points within the United States to ports. U.S. ports represent component two of the system, which involves the transfer of goods between the inland transportation system and oceangoing vessels. Component three involves the oceangoing vessel moving from a U.S. port to a foreign port. Component four involves the receiving port transferring cargo from the oceangoing vessel to an inland transportation system. And the final component involves the receiving country's inland transportation system delivering the cargo from the port to its final destination. The same five components are, of course, involved in coastwise traffic (between domestic ports) and are reversed in the case of imports.

CHANGES IN THE WORLD ECONOMY AND TRANSPORTATION SYSTEM

Concern with ports is intimately tied to the concern with international competitiveness which has become intense with the changing relationship between the U.S. and the world economy. Over the last two decades this change has accelerated rapidly as the U.S. economy has moved from the post-war period of satisfying domestic markets and supplying the world with a vast array of goods and commodities to one which is now the largest component of an increasingly interdependent world economy. For the first two decades following World War II, foreign trade was not a significant factor with regard to U.S. economic well-being. By 1980, however, 19 percent of the goods made in the United States were exported (up from 9 percent in 1970), and more than 22 percent of the goods consumed in the United States were imported (up from 9 percent in 1970). A perhaps even more telling statistic is that "by 1980 more than 70 percent of all goods produced in the United States were actively competing with foreign-made goods" (Reich, 1983). It is these economic changes that underpin an increasingly widely held belief that the future economic well-being of the United States is dependent on the nation's capacity to compete in a world economy.

There is a growing view that the United States must pursue every avenue in its effort to increase its competitiveness and therefore its exports. That is because exports are needed to pay for an increasingly high volume of imports. In 1984 the U.S. experienced its largest-ever trade imbalance with imports exceeding exports by more than $100 billion. Two options are frequently identified for addressing this trade imbalance. One is to restrict imports into the United States by using public policy to build barriers to those imports. As the trade imbalance has grown, increasing pressures have developed for taking such restrictive policy action. The other proposal is to increase the competitiveness of U.S. exports such that those exports are capable of paying for imports. Free international trade, and therefore, opposition to building barriers to imports has characterized every Administration since the end of World War II. It is widely believed, however, that unless the United States can become more competitive, the pressures for restricting imports will become irresistible. It is in this context, then, that the present and future adequacy of U.S. ports has become an integral part of the broader debate over the future competitiveness of the United States in the world economy.

Key to understanding the changing relationship of the U.S. to the world economy is an appreciation of the changing character of this nation's imports and exports. Reich (1983) suggests some of the changes: "During the 1970s the share of American manufactured goods in total world sales declined by 23 percent while every other industrialized nation except Britain maintained or expanded its share. American's diminishing presence in the international market has been particularly marked in capital-intensive, high-volume industries. Since 1963, the U.S. proportion of world automobile sales has declined by almost one-third. United States sales of industrial machinery also declined by one-third; sales of agricultural machines by 45 percent; telecommunications machinery by 50 percent; metalworking machinery by 55 percent."

In the period immediately following World War II when the industrial capacity of Europe and Japan was being rebuilt, the United States experienced an export boom and supplied some 60 percent of the world's manufactured goods. As Europe and Japan regained industrial capacity, trade between the United States and these other areas of the world moved into relative balance. Within the last decade, the flow of mass-produced industrial goods has reversed with the United States becoming a major importer and Europe and the Pacific Rim countries becoming major exporters.

In the present period, U.S. exports have come to be dominated by such bulk commodities as coal, grain, and timber and by what are now regularly characterized as high-technology products, plus sophisticated services such as communications and computer software.

Highlights of the changing character of U.S. trade in the past 50 years are presented in the following table. In 1947, as indicated on the table, the United States had a $9.5 billion trade surplus, as the nation supplied the world.

Highlights of U.S. Export and Import Trade: Exports Minus Imports (in millions of dollars)

Year	Agri- cultural Goods	Fuels and Lu- bricants	Chemi- cals	Capi- tal Goods	Con- sumer Goods	Auto- motive Products	Mili- tary Goods	Other	Total
1930	-459	433	3	518	-92	282	7	-271	782
1937	-459	395	22	486	-38	353	22	-184	265
1947	1604	1013	553	3144	958	1147	174	890	9530
1960	857	-739	1128	4949	-505	633	804	-1226	5528
1970	558	-1384	2216	10557	-4834	-2242	1230	-3163	3303
1973	8023	-6369	3137	13928	-8481	-4543	1385	-5854	1863
1981	24308	-71333	11995	45680	-22864	-11750	3608	-11325	-27566

SOURCE: Branson, 1984.

This early post-World War II boom utilized ports built and expanded during the 1930s, when an extensive program of public works was undertaken to provide employment in response to the Depression and to establish the infrastructure for regional economic development. Many of the dredged navigational facilities authorized for ports in the 1930s and 1940s are still being maintained as they were created then. The vessels that came into service following World War II were the war-surplus "handy-size" tanker, and the Liberty and Victory general cargo ships. These small, flexible vessels were nicely accommodated by U.S. ports.

In the 1960s, as industrial capacity recovered elsewhere, world trade grew rapidly. The economic activities of all nations benefitted from inexpensive Middle Eastern oil. To serve expanding world trade, shipowners began building a different type of fleet. Ships became increasingly larger and more specialized; for example, supertankers.

As the United States accelerated its oil imports, the need was perceived for ports capable of serving supertankers. Efforts to develop these facilities became controversial when major oil spills occurred in the Santa Barbara Channel on the Pacific Coast and elsewhere in the world. After considerable controversy, however, Congress enacted the Deepwater Ports Act of 1974, which provided for the development of offshore petroleum facilities. One was built (Louisiana Offshore Oil Port, or LOOP) by five oil companies in 1979. Plans for other such facilities, however, were abandoned as the consequences of the oil disruptions of the 1970s were felt and Middle Eastern oil imports declined. A key factor, of course, was a nearly tenfold increase in oil prices. Worldwide, the transport of oil plummetted.

The oil shocks of the 1970s had equally serious consequences for the nation's trade balance. A few data suggest the implications of these developments. In 1970, imported oil cost the nation $2 billion. By 1974, the cost of these imports rose to $6.5 billion, and by 1980, oil imports cost the nation $78.9 billion. Estimates are that in 1984 oil imports will cost about $60 billion. Almost all projections suggest that a high oil import bill will continue.

Worldwide, one response to accelerating oil costs was to search for alternative energy sources. The most abundant, readily available alternative was coal and around the world, nations quickly sought to substitute coal for some of their oil imports. It was this rapid move to coal that focused the present national attention on the inadequacy of ports. With large, readily available coal reserves, the United States experienced a surge in demand for its steam coal in 1980. That demand was triggered by a combination of the Iranian oil disruption and unstable conditions in other major coal exporting nations. During 1980, newspapers in the United States were full of reports of large numbers of colliers waiting for weeks and sometimes months to gain access to U.S. coal-loading facilities.

A number of studies during this period (Energy Information Administration, 1981; ICF, Inc., 1981; National Coal Association, 1981; Wilson, 1980) concluded that the United States had an opportunity to become a major supplier of a massive new world market for steam coal. To gain and secure that market, however, it was repeatedly noted that the United States would need to be able to handle the most efficient dry or combination bulk carriers requiring water depths greater than those available in U.S. coal ports.

In the years since 1980, the development of a world oil surplus, the declining price of oil, a strong dollar, and the reestablishment of political stability in Poland and labor stability in Australia have reduced the demand for steam coal exports from the United States. Whether the present situation with regard to world energy will be sustained for a long period remains an open question. A major oil disruption in the Persian Gulf could trigger renewed demand for U.S. coal exports.

In combination, then, the changing role of the U.S. in the world economy, the changing character of U.S. exports and imports, and the unpredictable world energy situation have created substantial uncertainty with regard to future port capacity needs. Projecting the future size and character of world shipping is extremely difficult. As a result, few mid- and long-term shipping forecasts have been made in recent years. Lloyd's Register of Shipping (1984) suggests why such forecasts are not being made by noting: "World shipping and ship building are experiencing the worst economic recession in the last 50 years and the interaction of technical, commercial, and political factors makes it very difficult to predict the likely rate of recovery."

The hesitancy in making long-term forecasts is readily understood when one reviews the inaccuracy of forecasts made in the past (see Figure 1, Appendix G*). Estimates of liquid cargoes, principally oil, show the most striking contrast with actual cargoes carried.

---

*Tables and full-page figures have been placed in Appendix G for the convenience of the reader.

While the total volume of goods and commodities carried in
oceanborne trade follows the world economy, the relationship that
previously prevailed between world GNP growth and growth in demand for
oceanborne transport was disrupted in the 1970s. In the following
table, oceanborne transport is divided into three categories: oil,
bulk, and other. From 1965 to 1973, the growth in the quantity of oil
transported annually averaged a rate that was more than twice the rate
of growth of the world GNP. From 1973 through 1980, the quantity of
oil being transported remained almost stable while world GNP was
growing at an annual rate of 2.3 percent. In the case of bulk
commodities, the quantities being transported continued to grow more
rapidly than GNP but less rapidly than during the 1965-1973 period.
By comparison, during the 1973-1980 period, the quantities of other
commodities in ocean transport grew at a rate, when compared to world
GNP, that was rougly twice that of the 1965-1973 period.

Annual Average Increase (%), World GNP and Oceanborne Freight
Transportation

Period	World GNP	Total		Oil		Bulk		Other	
		Tons	Ton-Miles	Tons	Ton-Miles	Tons	Ton-Miles	Tons	Ton-Miles
1965-1973	4.6	8.4	12.9	10.5	16.0	7.9	10.2	5.2	6.4
1973-1980	2.3	2.2	1.2	0.2	-1.0	3.0	3.9	4.9	5.7

aCoal, grain, bauxite/alumina, iron ore, rock phosphate
SOURCE: Maritime Transport Committee, 1981.

One result of these changes and optimistic forecasts was that
vessels ordered during the period of growth became surplus. For the
first time since World War II, during both 1982 and 1983, the total
deadweight of the world fleet declined, yet the rate of scrapping was
insufficient to bring cargo-carrying capacity into balance with
available cargoes.
Nonetheless, shipowners continued to order vessels in 1983, "as
covert and overt subsidies [encouraged] owners to replace aging
vessels,...in addition, the [shipbuilding] industry was offering more
efficiently designed ships with the emphasis on fuel economy, and
finance was freely available" (Lloyd's Register of Shipping, 1984).
A particular emphasis has been on containerships which are
typically employed in the liner trades. The growth of trade in goods
that can be packed and shipped in containers offers partial
explanation. The number of containerships increased 5.7 percent in
1983; their container-carrying capacity increased 7.5 percent
(Maritime Transport Committee, 1984).
Table 1 (Appendix G) indicates the variety of vessels in the world
fleet that called on ports of the United States in 1980 (engaging in
foreign trade). Considerable specialization of vessel types can be
seen in this list, much of it matching the changing mix of U.S.
imports and exports previously discussed. Design drafts for a number

of the bulk carriers exceed the water depths of many U.S. ports. For
bulk carriers with more than 46 feet of draft, water-depth limitations
would prevent their being fully loaded, incoming or outgoing, in most
of the major bulk-commodity ports in the United States (except Los
Angeles, Long Beach, an oil terminal in Bellingham, Washington, or
grain terminals in Seattle and Tacoma).

SUMMARY

What can one conclude from this mix of changing and sometimes
conflicting data? The changing relationship of the U.S. to the world
economy and the changing character of the world's commercial fleet
make confident projections of future shipping patterns extremely
difficult. So far as the needs of U.S. ports are concerned, and
specifically the need for new construction dredging, the picture is
characterized by great uncertainty. That uncertainty has doubtless
been a factor contributing to the stalemate in new port construction
activities.

PORT DECISION MAKING IN THE UNITED STATES

The second factor contributing to the U.S. port construction stalemate
has been what was previously characterized as the unraveling of the
social contract between the federal government and the nation's ports
concerning funding, management, and regulation of dredging. To
appreciate what has occurred, a brief review of the evolution of that
social contract is useful.

The beginning of this history occurred in the very early years of
the Republic. Prior to 1824, river and harbor improvements were
commonly executed and paid for by state and local governments. In
this early period, federal responsibility covered navigation and
safety services such as coastal charts, lighthouses, and beacons.
Congress authorized states and individual ports to levy tonnage duties
to pay for such work (Hill, 1957).

Direct federal involvement in port construction and maintenance,
specifically dredging, was initiated in 1824 with the passage of the
General Survey Act. Under this legislation, Congress made its first
appropriations for rivers and harbors improvements. Since that time,
the federal role in port construction and maintenance has been
inextricably tied to the development of the U.S. Army Corps of
Engineers. The Corps was initially chosen because of its unique
engineering expertise. President James Monroe advocated that the
Corps serve as the national planning organization for rivers and
harbors. Congress, however, rejected the notion of any national
planning responsibility on the part of an executive agency and
established a pattern of deciding and funding port developments on a
case-by-case basis.

The rejection by Congress of President Monroe's proposal for a
national planning role for the Corps and the decision to authorize and

fund port projects on a case-by-case basis deserves special emphasis. It established a detailed decision-making role for Congress (and the Corps) which has meant that for more than 150 years, the Corps has carried out the mandates of Congress with Presidents having only very limited oversight and management control of these activities.

This special relationship between the Corps and Congress has been the object of criticism off and on for the last 150 years. The nature of that criticism has remained essentially the same. Examples of the criticism are as follows: The system provides no national plan for ports and makes no distinction between ports of national versus local value; the process is dominated by logrolling and pork barrel tradeoffs; the system reflects sectional favoritism; and the system funds many projects that cannot be justified on an economic basis (Hill, 1957). As early as 1830, President Andrew Jackson pocket-vetoed a rivers and harbors bill because it did not distinguish between works of national and those of local value. If the criticism of the system that developed has remained consistent, so has the basic framework of the relationships among Congress, the Corps, the Office of the President, and a range of local and national interest groups. The most distinctive characteristic of the system is the level of detailed control that Congress exercises with regard to Corps projects generally and port projects specifically. Corps projects are distinctive in that Congress provides year-to-year funding for multiyear construction projects.

This year-to-year funding is in contrast to the more typical full-funding approach, which characterizes major construction projects carried out by most other federal executive agencies. Under the full-funding approach Congress includes the entire cost for multiyear projects in a single annual budget. Full funding allows the executive agency much more authority and discretion than year-to-year funding (Scheppach, 1977). Specifically, full funding allows the executive agency to reprogram funds from one project to another independent of specific Congressional approval. Second, multiyear funding gives the White House, through the Office of Management and Budget, substantially more control over the agency than is the case with the year-to-year funding that characterizes Corps projects.

Alternatively, the project-specific, year-by-year funding approach results in a uniquely tight budgetary relationship between Congress and the Corps and gives the Corps a great deal of independence from the normal executive budget process (See Ferejohn, 1974; Maass, 1951; and Hill, 1957). From the congressional side, this funding approach allows Congress and the specific congressmen and senators concerned with individual projects a great deal of control. Key to the success of a project is the capacity of the congressmen interested in specific projects to negotiate with their peers in a process that essentially involves trading support for each other's public works activities. Individual ports, then, develop tight links both to the local Corps districts and to their congressmen in promoting new construction. These represent micropolitical systems organized around individual ports.

Over the course of nearly a century and a half, the decision making process in Congress associated with the rivers and harbors legislation demonstrated a capacity to respond to perceived port needs. The history of this decision making process is one that has involved swings between periods of growth and periods when few new public works were undertaken. Viewed historically, however, this decision making system has generally proved satisfactory for those interested in port development.

Beginning in the early 1970s, however, it was no longer possible for Congress to evolve decisions which allowed for the initiation of major, new, federally funded port projects. Several factors have been suggested as major contributors to this stalemate, but three are regularly identified. They are: (1) broad public concern with environmental consequences, (2) rising budgetary deficits, and (3) basic changes in public attitudes toward major federal construction projects.

During the 1970s, broad public concern developed with environmental protection. In response to this public concern, Congress passed several pieces of environmental legislation mandating a variety of federal agencies to put environmental regulatory programs in place. In the case of port construction projects, the Corps was assigned responsibility for assessing the environmental consequences of port projects and assuring through a complex approval process that environmental concerns would be an integral part of the decisions made. As a part of this development, local citizens groups and a variety of state and federal agencies with environmental responsibilities became active participants in the decision making associated with port construction. Two consequences resulted from these developments. First, the process of port construction became significantly more complex, and second, the time required to meet these environmental responsibilities extended the period and cost required to carry out major port projects.

Also, beginning in the 1970s, public concern with ever larger deficits increased. No longer able to count on getting their fair share of an ever-expanding federal budget, political leaders and their constituencies faced difficult tradeoffs among programs and projects. Political tradeoffs have always shaped the nation's policies and certainly its choice of public works projects. But by the late 1970s and the early 1980s, there existed such a mismatch between the public's expectations for government services and the government's fiscal capacity to deliver them that stalemate began to characterize many areas of public policy. (See Levine, 1980, for a collection of essays on the financial crisis in the public sector.)

As deficits grew, so-called discretionary federal funding became the focus of intense attention. Many saw the omnibus water-resources projects bill (for all water projects funded by the federal government, including port dredging) as containing the most discretionary of federal expenditures. It should be noted that one of the characteristics of these public works expenditures is that all it takes to contain such expenditures is inaction. The annual funding approach reflected in the rivers and harbors legislation, therefore, required on the part of those congressmen opposing public works expenditures only that they refuse to join a consensus in funding authorized projects.

At a more general level, during the 1970s, there were increasing calls for a brake on big government. Although these calls for reducing the size of government meant different things to different people, a common theme was opposition to special interest projects viewed as being uneconomic. During the Carter Administration, there was particularly strong opposition in the White House to Western water projects. These projects were regularly criticized as uneconomic and inefficient expenditures of federal funds. Since port dredging is handled in the same legislation as these and other public works to develop water resources, dredging projects were also opposed. Remember that the original rationale for handling diverse public works projects in a single piece of legislation was the desire to utilize the engineering expertise of the Corps. Recall also that criticism of the local and uneconomic character of many public works projects goes back to the early 19th century. Present-day critics frequently see no distinction between port construction and the construction of dams and other water projects. The case-by-case authorization and annual funding approach in combination with the lack of any enabling legislation or broadly stated national policy which provides criteria for distinguishing between projects of local and national value, left the nation, by the 1980s, without an established framework for setting priorities among public works projects--as, for example, between dams and ports.

Many of the proponents of additional port construction argue that there is a fundamental difference between a major port that links the United States to the world economy, an economy on which the nation is increasingly interdependent, and a dam serving a local area of the United States. These proponents emphasize that major ports are clearly of great importance nationally, while many other public works projects are, in fact, primarily in the local or regional interest.

In the face of changing attitudes toward federal public works projects, a growing federal deficit, and broadly based public concern about environmental values, the system for making decisions about port construction that had evolved incrementally over the history of the American Republic was becalmed. Given the belief that additional port capacity is essential to the economic well-being of the United States and at the same time opposing additional large public expenditures, the Reagan Administration early in its first term sought to break the logjam on ports by proposing establishment of a port user fee. The rationale behind the Reagan Administration's user fee proposal was that it would allow nationally important port construction to be undertaken, and ensure equity and efficiency. That is, those who benefit pay, thus equity is achieved; only those projects that can pay their own way are carried out, thus efficiency is achieved.

In seeking to understand the nation's port needs, the impediments to achieving those needs and the options for dealing with those impediments, this report investigates a broad set of concerns. These range from economic and engineering concerns through institutional and environmental concerns to the general character of worldwide oceanborne shipping. Conceptually, this report starts with the most general questions. Although additional construction and maintenance dredging has powerful advocates, it is important to note that there are opponents who have concluded that new construction dredging is unnecessary. The crux of the debate over dredging needs revolves around the costs and benefits of the additional investments that would be required for U.S. ports to be able to handle ships of larger size.

This report, then, moves in Chapter 4 to an investigation of the relative advantages and disadvantages to the United States of being able to handle larger-volume ships. In Chapter 5, the report investigates the relative attractiveness of additional dredging in existing ports versus a variety of other ways of handling large-volume ships. Chapter 6 considers the various proposed approaches to funding federal projects and their implications for the over-all port construction and maintenance system. Chapter 7 describes the institutional decision making system for dredging and ways of bringing stability, predictability, and speed to that decision making. Chapter 8 addresses technical needs and issues associated with both new construction and maintenance dredging. Chapter 9 assesses the environmental issues associated with dredging, the state of scientific knowledge with regard to these issues, and some approaches to addressing the issues.

REFERENCES

Branson, W. H. (1984), "Trade and Structural Adjustment in the U.S. Economy: Response to International Competition," Discussion Paper #70, Woodrow Wilson School, Princeton University, Princeton, N.J.

Democratic National Committee (1983), "Response to the State of the Union Address," Broadcast, Democratic National Committee, Washington, D.C.

Energy Information Administration (1981), "Interim Report of the Interagency Coal Export Task Force," Draft, Washington, D.C., U.S. Department of Energy.

Ferejohn, J. A. (1974), Pork Barrel Politics, Rivers and Harbors Legislation, 1947-1968 (Stanford, Calif.: Stanford University Press).

Fearnley's (1982), World Bulk Trades 1981 (Oslo, Norway: Fearnley's).

Hill, F. G. (1957), Roads, Rails, and Waterways (Norman, Oklahoma: University of Oklahoma Press).

ICF, Inc. (1981), Potential Role of Appalachian Producers in the Steam Coal Export Market (Washington, D.C.: ICF, Inc.).

Levine, C. H. (1980), Managing Fiscal Stress: The Crisis in the Public Sector (Chatham, N.J.: Chatham House Publishers).

Lloyd's Register of Shipping (1984) Annual Report 1983 (London: Lloyd's Register of Shipping).

Lloyd's Register of Shipping (1983), Statistical Tables (London: Lloyd's Register of Shipping).

Maritime Transport Committee (1984), Maritime Transport 1984 (Paris: Organisation for Economic Cooperation and Development).

25

Maass, A. (1951), <u>Muddy Waters: The Army Engineers and the Nation's Rivers</u> (Cambridge, Mass.: Harvard University Press).

Maritime Transport Committee (1984), <u>Maritime Transport 1984</u> (Paris: Organization for Economic Cooperation and Development).

Maritime Transport Committee (1983), <u>Maritime Transport 1982</u> (Paris: Organisation for Economic Cooperation and Development).

Maritime Transport Committee (1982), <u>Maritime Transport 1981</u> (Paris: Organisation for Economic Cooperation and Economic Development).

National Coal Association (1981), "A Forecast for U.S. Coal in the 1980," Washington, D.C., National Coal Association.

Reich, R. B. (1983), <u>The Next American Frontier</u> (New York: Times Books).

Scheppach, R. D. (1977), "Water Resource Funding--A Congressional Perspective," Paper presented at the Annual Meeting of the American Economics Association, December, New York City.

Schonknecht et al. (1983), <u>Ships and Shipping of Tomorrow</u> (Centreville, Md.: Cornell Maritime Press).

Wilson, C. L. (1980), <u>Future Coal Prospects: Country and Regional Assessments--Report of the World Coal Study</u> (Cambridge, Mass.: Ballinger Publishing Co.).

4
Does the United States Need to Accommodate Large Vessels?

INTRODUCTION

Assessing the nation's need for additional capacity to accommodate large vessels involves balancing a complex set of factors. Proponents of additional capacity give three reasons: (1) economics, (2) national security and defense, and (3) the need to be able to respond rapidly and flexibly to future changes in the character of the ocean transportation system. This chapter appraises the purported need for additional capacity to handle large vessels in terms of these three categories of justification.

As an initial step, it is necessary to define "large ships." Two categories of deficiency in port capacity have been identified in studies conducted by the U.S. Army Corps of Engineers and by the ports of the United States: (1) the limited ability of the United States to handle large bulk carriers, and (2) the need by some ports to handle medium size vessels; in particular, the latest-generation containerships, but also other specialized or general vessel types of Panamax dimensions. The limits of the Panama canal are 900 ft length, 106 ft beam and 42 ft draft (draft limits vary with water supplied to the canal and may in some seasons be less). These two identified needs have been taken by the committee as defining "large ships."

Proposals for additional dredged capacity are briefly summarized in Table 2 (Appendix G). The range of large bulk carriers cited in these proposals have cargo-carrying capacities of 105,000 DWT to 150,000 DWT. The latest-generation containerships referred to vary in length from 800 ft to 950 ft, and in beam between 105 ft and 110 ft. These vessels have cargo-carrying capacities of about 3500 TEUs* and require water depths of 37 to 43 ft. Owing to the windage area of these vessels when containers are loaded on deck, additional channel width may be necesary, particularly in bends or turns.

This chapter discusses the present port situation in the United States for accommodating large vessels, status and trends in the world

---

*A TEU is a "twenty-foot-equivalent unit." Containers are 8 ft wide, and 10, 20, 30, or 40 ft long. "TEU" is the standard for comparing container-carrying capacity.

fleet, arguments that have been advanced for and against creating the capacity to handle large ships, the present situation of other ports worldwide, and considerations of national security and defense.

WHAT IS THE PRESENT PORT SITUATION?

It has been suggested that the United States is already paying penalties because of its limited capacity to accommodate large ships; specifically, that many large vessels call on the ports of the United States less than fully loaded, owing to the limited water depths in navigational channels. A review of 1980 data indicates that 3849 port calls were made in the United States by vessels with design drafts greater than 45 ft, but only 754 of these port calls were at actual drafts greater than 45 ft. For most of the major ports of the United States, water depths become limiting for vessels at about 40 ft of draft. Additional feet of draft are achieved by operating vessels in navigational channels at slower speeds, by taking fully loaded vessels in or out of port at high water, and other means, but the additional draft that can be achieved by these means is limited.

While the inference might be drawn that light-loading is always due to channel limitations, there are several other reasons. A major one is avoidance of a ballast leg; for example, a large combination (liquid and dry bulk) carrier may deliver a large oil cargo to a U.S. port and load a smaller grain cargo at another U.S. port for export. Other reasons are discussed in succeeding sections.

Channel depths at mean low water for the coastal ports of the United States are listed in Table 3 (Appendix G), together with the number of port calls at vessel drafts at or exceeding channel depths (at mean low water) in 1981.

For the vessels defined in this report as large ships, data for 1980 indicate that 476 vessels greater than 90,000 DWT made 2863 port calls in the United States. Bulk carriers (liquid, dry, combination) that would have drafts of 46 ft or more fully loaded made 1590 port calls that year, but only 236 were at actual drafts of 46 ft or more. Of these, 26 were at the deep-water terminals of Long Beach, Los Angeles, or Puget Sound. Thus, less than 13 percent of total port calls by bulk carriers were at 46 ft or more in draft. Table 4 (Appendix G) presents port calls in 1980 by large liquid- and dry-bulk carriers at the four ports with approved plans for construction dredging to accommodate such vessels. While definite conclusions cannot be drawn from the table, shippers and ship operators dealing in bulk commodities in these and other ports uniformly emphasized to the committee their need to load additional feet of draft. Draft limitations for tankers or bulk carriers limit their cargo-carrying capacity as follows: for a 150,000 DWT or 200,000 DWT vessel, failure to use a foot of draft may translate into 3400 to 4900 long tons of lost cargo-carrying capacity. Some representative figures are given in Table 5 (Appendix G).

Table 6 (Appendix G) lists U.S. port calls by containerships in 1980, in comparison with port calls by other vessels carrying general cargo. Owing to the aggregation of the data from this source into a large category of 16 ft to 45 ft of draft, it is not possible to determine the use being made of U.S. ports by large containerships. Delivery has been taken in the past three years of several large containerships, and some have called on U.S. ports. Great care has been exercised in some ports to achieve transit of these vessels, such as entering/leaving at high water, one-way traffic, and operation at slow speeds. If in the future two-way traffic, higher speeds, or other operational changes are desirable, greater channel dimensions may be required. Efficiency or higher traffic density may depend on depth. If vessels pass or overtake one another, or if higher speeds in port are desirable, or if shipowners want to be able to enter and leave port at periods of low water, these operations require more depth (and possibly greater widths) than one-way traffic, slower speeds, and transits at high water (see Chapter 7).

Thus, the broad implication that can be drawn from these summary data is that the physical dimensions of navigational channels in the United States are being used in some instances to capacity (and beyond). Whether significant benefits are to be gained (and if so, how much and over what period) from additional navigational capacity is difficult to determine, owing to the volatility of trade, and of oceanborne shipping, and several uncertainties affecting projections of the future, as discussed in succeeding sections.

Vessels in the World Fleet and Major Trends

Tables 7, 8, and 9 (all in Appendix G) list the dry bulk and combination carriers, tankers, containerships and roll-on/roll-off vessels (Ro-Ros)* of the world fleet. Inspection of Tables 7 and 8 indicates that the proposals of U.S. ports to increase their capacity to handle large bulk carriers are not for the largest of these vessels. The proposals to dredge deeper channels for 105,000 DWT to 150,000 DWT vessels encompass perhaps 600 vessels in the world fleet, but note that because of variation in draft in this range of cargo-carrying vessels approximately 300 to 350 of the vessels may still need to be light-loaded by 1 ft to 5 ft to call on U.S. ports even after deepening. There remain beyond these vessels 82 dry bulk carriers and 475 tankers that are larger, and that would be significantly light-loaded if they called on the deepened ports.

Containerships in the world fleet and on order can be categorized in two groups: those having less than 30 ft of draft (79 percent), and much larger vessels with container-carrying capacities of 3000 or 3500 TEUs or more, and design drafts of 37 to 43 ft. It should be pointed out here that there is greater ambiguity between the design and actual draft of a containership (fully loaded) than is the case with bulk carriers and tankers carrying dense and fairly uniform

---

*Roll-on/roll-off vessels are equipped with ramps, over which their cargoes are loaded or unloaded. Some also carry containers.

cargoes. Tankers and bulk carriers tend to be weight-limited and containerships tend to be volume-limited. Thus, tankers and bulk carriers fully loaded with dense cargoes will be at or near their maximum draft (as indicated by load lines), but containerships loaded with the maximum number of containers they were designed to carry rarely approach maximum draft.

With the great number of smaller vessels in the world fleet, why are the larger vessels of concern to ports? First, many of the vessels included in summary tables of the world fleet are designed for short sea routes (as, for example, between European ports) and coastwise trade.

Second, the maximum achievable speed of vessels is related by physical laws to their shape and length. Since the increase in the price of fuel, speed requirements have been reduced, but any change in the future giving an advantage to speed will favor larger vessels.

Third, the transportation costs per ton of cargo (or per container) are lower for larger than for smaller vessels, as illustrated in Figure 2 (Appendix G).

Primarily because of their lower transportation costs and higher productivity, movement to larger vessels has been the trend for the past two decades, as illustrated in Figure 3 (Appendix G). This trend was first evidenced for tankers, and although the lack of deep water in the ports of the United States and many other oil-importing countries has resulted in the use of smaller tankers to lighter larger tankers, 72 percent of the world's oil supplies are carried in tankers 100,000 DWT and larger (Cargo Systems Research Consultants, 1982). Bulk carriers 100,000 DWT or more increased their share of oceanborne shipments from 6 percent in 1971 to 35 percent in 1980 (H. P. Drewry, 1982). Large bulk vessels dominate iron ore trade: 80 percent of iron ore shipments were carried in vessels 100,000 or more in 1981. The trend to larger vessels for iron ore and coal was reinforced by the introduction of combination carriers (oil, ore, dry bulk), which tend to be larger than dedicated dry-bulk carriers. Large bulk carriers now carry 45 percent of all coal shipments, and 10 percent of grain shipments. Other bulk cargoes (the "minor bulks"--phosphate rock, sulfur, wood chips, etc.) are carried in smaller vessels. Among the commodities imported or exported in vessels calling on U.S. ports at 46 ft of draft or more in 1980 were corn, edible oils, sugar, iron ore, sulfur, chemicals, rubber, coal, oil and oil products, and vehicles.

Containerships, because of their much higher productivity as compared to other general cargo ships, are displacing older general cargo vessels, and their productivity appears to increase with size. Cargo-carrying Ro-Ros of large size are also growing as a percentage of the world fleet, as they are flexible vessels (many carry containers) able to load and unload in a great variety of ports, and able to carry cargoes too large or awkward for packing into containers. In the world fleet, both types of vessels are of recent vintage--80 to 90 percent are less than 10 years old. Aggressive building programs have been instituted the past three years aimed at replacing smaller with larger containerships and Ro-Ros. The latest-generation containerships are entering round-the-world service in attempts by large liner companies to retain and enlarge their share of markets (Maritime Transport Committee, 1984).

There are several possible difficulties in assessing the reasons for the trend to larger ships. One is that while larger ships offer economies of scale in transportation, they may or may not have higher loading/unloading costs in port.

Another difficulty is that economies of scale are usually represented in terms of the costs, rather than the actual price, of oceanborne transportation. For example, the market prices of oceanborne shipping in bulk carriers for the past three years, while reflecting lower costs for transport in larger vessels, also reflect a smaller price spread between transport in larger and smaller vessels (some representative prices are given in Table 10, Appendix G) than the cost differential would suggest. Price is set by the perceived relationship of supply and demand for various vessel types, and the past three years have been characterized by low (uneconomic) prices for transportation in all bulk carriers. Owing to the high rate of new orders for bulk carriers and (despite record scrapping) continued surplus of tankers, prices may remain depressed for some time into the future.* In sum, price differentials do not necessarily reflect cost differentials.

Other difficulties make shipping hard to assess. While shipowners would like always to sail fully loaded, the size of cargoes may be determined by shippers in a surplus market. For the past three years, trade in the major bulk commodities has been characterized by spot markets, rather than long-term, fixed contracts, and the same has been true for vessel charters. During the recent worldwide economic recession, demand for vessels was influenced by the use of existing stocks of major bulk commodities, and the amounts shipped tended to be smaller than the amounts consumed (Maritime Transport Committee, 1984). Other factors which can influence port needs are suggested by a new trade pattern that emerged for coal in 1982. It illustrates some of the uncertainties associated with projecting the additional benefits to be gained from channel improvements. Large bulk carriers (100,000 DWT and more) were light-loaded with coal in Atlantic ports in the United States and sailed to Japan via the Cape of Good Hope, where they were topped off in a deep coal port (Richards Bay, South Africa). It is not possible to predict whether deepening the coal ports of the United States would capture this additional amount or whether buyers will still prefer to buy from both sources. Yet another possibility is that even larger vessels might be used, loading to 150,000 DWT or so in the United States and topping off in South Africa or elsewhere.

---

*Some observers are pessimistic about a near-term balance of supply and demand in these vessels and in liquid and dry bulk cargoes (Maritime Transport Committee, 1984), others are optimistic that balance will be achieved in a few years (Office of Technology Assessment, 1983).

Understanding why light-loading occurs requires information about the maximum desired sizes of shipments, frequency of delivery, amount of stockpile desired by various customers, time-value of the stockpile, and the influence of political decisions and new technology on demand.

Political considerations are sometimes significant in decisions determining the composition and characteristics of the world fleet and in its deployment. For example, governments subsidize shipyards for noneconomic reasons and give preference to their ships when the government is the customer. Many newly industrializing nations that are seeking to build a merchant marine protect their fleets by assuring those fleets a share of the nation's imports and exports. These considerations may result in agreements reserving cargoes for national-flag fleets and other stipulations that will (besides inhibiting competition) enhance or reduce the need to accommodate large vessels in the future.

Ports Worldwide

Ports in other maritime nations have perceived a need to increase their capacity to accommodate large vessels. There are now 76 ports worldwide with depths greater than 55 ft (Table 11, Appendix G). Most of these ports export or import one (or more) major bulk commodity. To gain information about the expectations of world ports, the committee sent a list of questions (Appendix D) to a large number of ports with the assistance of the International Association of Ports and Harbors and the embassies of several maritime nations.

Of the 59 ports responding, 22 indicated they had plans for expansion of navigational facilities, or that expansion was under way, and 5 had just completed improvements. The responses of these ports are briefly summarized in Table 12 (Appendix G).

Some ports that are already between 55 ft and 64 ft deep are planning further improvements--Antwerp and Zeebrugge, Belgium, for example. Richards Bay, South Africa, has just completed deepening to 64 ft, and is now deepening to 75.9 ft. Two of the ports in the very deep category (greater than 65 ft) noted the need for a new deep-water port in their respective countries (Mombasa, Kenya, indicated plans for a deep water port at Lamu; Su-Ao, China, stated that another deep-water port was needed).

Review of the existing and planned dimensions of ports elsewhere in the world suggests that a large number expect to need the ability to handle large vessels in the future.

ECONOMICS

Clearly, the primary factor influencing the movement toward larger ships is that they offer lower transportation costs. Stated in the simplest terms, the argument for additional channel dimensions to handle larger ships is to allow the nation to enjoy the transportation savings that ultimately flow from using larger ships. In the absence of adequate port capacity, large ships must either enter and leave U.S. ports less than fully loaded or U.S. trade must be carried in smaller ships. In either case, the cost to the nation will be higher.

Most of the controversy over whether there is a need to develop additional port capacity to handle large ships revolves around whether the reduced transportation costs flowing from that capacity will be sufficient to cover the costs of developing the capacity. In general, the debate has been focused on proposals for deepening existing ports. Table 17 (Appendix G) indicates the range of costs associated with various proposals for port deepening. In the case of those projects calling for deepening to 55 ft, construction estimates range from $372 million to $440 million. In each of these instances, the U.S. Army Corps of Engineers' analysis of benefits and costs concludes that the benefits outweigh the costs. Many factors influence whether benefit/cost analyses turn out to be positive or negative. The most critical single variable is the expected level of use. Future use of additional port capacity, however, is inherently uncertain.

The committee posed two questions for itself: Is there sufficient present need to justify developing additional capacity? Will there be sufficient future need to justify developing additional capacity?

With regard to the first question, the committee could find no convincing evidence of present needs to develop additional capacity to handle large ships. It should be emphasized, however, that even assessing present needs involves dealing with considerable uncertainty. Perhaps the key example was the committee's inability to determine whether (and how much) channel limitations result in the use of smaller rather than larger vessels, or light-loading of large vessels.

The level of uncertainty is substantially greater in answering the second question; that is, future needs. Confident assessment of future port needs would require reliable information on: (1) the over-all size and growth patterns of the future world economy, (2) the level of oceanborne U.S. exports and imports, and (3) the character or mix of those exports and imports.

Data presented in Chapter 3 suggest two trends with regard to the U.S. and world economy. First, the United States is increasingly becoming an interdependent part of the world economy. Among important U.S. exports to industrializing countries, for example, have been (and are) manufacturing machinery and equipment, while manufactured goods produced at lower cost in those countries are imported into the U.S. The U.S. both exports and imports raw materials and agricultural products. These relationships involve trade among countries in various geographical regions. The growth of manufacturing capacity in many countries is broadening and diversifying patterns of trade. Low-cost oceanborne transportation is generally seen as contributing to the enhancement of interdependent national economies and to the promotion of world trade which ultimately benefits the United States. But while the world economy and world trade are expected to exhibit over-all growth in the mid- and long-term future, trade is volatile, and oceanborne transportation is more volatile. Thus, sharp, short-term fluctuations can be expected.

The primary pressure driving the demand for U.S. capacity to handle large bulk carriers rests on the expectation that there will be opportunities for increased exports of bulk commodities, particularly coal and grain. The potential world market for U.S. coal has been the most frequently used rationale for developing additional port capacity. The landed price of coal in foreign markets can be heavily influenced by transportation costs. For example, one recent study of the potential advantages of using large bulk carriers for transporting coal between Hampton Roads and Rotterdam indicated a potential cost per ton differential ranging from $10.33 to $6.95 (Graves et al., 1984). Depending on the assumptions used, these investigators found that transportation costs could vary by as much as $3.38 per ton. Assuming a landed cost for coal in Rotterdam of $50 per ton, these differential transportation costs could range from 14 to 20 percent of landed costs. Many analysts believe that this 6 percent differential in delivered costs could, for a low-value, high-volume commodity such as coal, make the difference between U.S. competitiveness and lack of competitiveness. Stated differently, these transportation costs could significantly influence the share of the world coal market supplied by the United States.

The authors of the study referenced here found compelling reasons for recommending caution in funding the dredging of coal ports. They did, however, find that, given optimistic assumptions about future European demand for U.S. coal and accepting what they referred to as a lower societal return-on-investment, dredging one coal port could be justified. In general, economic studies have recommended caution (Energy Information Administration, 1983; General Accounting Office, 1983).

The potential benefits to the U.S. grain export trade of being able to handle large bulk carriers is even less clear. First, because grain has a much higher value per ton, transportation costs represent a smaller portion of landed costs. Second, many of the markets for U.S. grain do not have adequate unloading facilities or ports with sufficient depth to handle large bulk carriers. These two points are regularly made. The committee would caution, however, that it was repeatedly informed of instances where large bulk carriers were used to transport grain. Additionally, the Soviet Union has recently been investigating the possibility of topping off grain ships in the Gulf of Mexico. We include these references simply to indicate that even in the case of grain, there is evidence that if additional port capacity were available, larger portions of U.S. grain exports might be carried in large bulk carriers at lower prices.

The relative advantages and disadvantages of being able to accommodate larger oil tankers are also difficult to evaluate. Oil is a flexible commodity that can be loaded and unloaded by several different technologies (as described in Chapter 5). There are additional costs associated with some of the alternatives, and the cost of oil imports might also be reduced by the ability to accommodate larger tankers in port.

The potential economic benefits of increased capacity to handle high-value cargo in larger ships have not been as extensively debated

or studied as those of increased capacity to handle bulk carriers.
The movement toward large containerships suggests economic advantages,
however, and the costs of developing the additional capacity to handle
these ships in some of the major U.S. ports may be economically
justifiable.

Recent history shows a significant increase in the quantity of bulk
commodities exported from the United States. A more narrowly focused
review of those data indicates another important characteristic. That
is, in the case of both coal and grain exports, the quantity of
exports can differ substantially from one year to the next. For
example, coal exports in 1981 amounted to 110.2 million tons. By
1982, those exports had dropped to 105.2 tons, and in 1983 to 76.9
million tons. Similarly, in 1978, the peak year for grain exports,
the U.S. sold 105.2 million metric tons overseas. By 1982, those
exports had dropped to 97.2 million metric tons.

The point which appears to deserve emphasis is that in bulk
commodities, extreme swings can occur over very short periods of
time. To take advantage of rapidly expanding markets, the nation
would benefit from having available port capacity when the swing is
upward. Because the buyers of these commodities are often concerned
with the delivered price, inability to accommodate optimal vessel
sizes and types may affect the nation's ability to secure long-term
agreements that at least minimize the freight rate (H. P. Drewry,
1981).

One conclusion seems demanded from a review of the preceding data:
it is that the United States faces great uncertainty with regard to
the size and character of the future world economy, the nature of
future oceanborne transportation into and out of U.S. ports, and the
future mix of commodities that the nation will export and import.
Further, the character of U.S. exports and imports, particularly
exports of such bulk commodities as coal and agricultural products, is
likely to change from year to year. In conclusion, the size and
character of future U.S.-world trade and oceanborne transport is
simply not now predictable over any length of time with any degree of
reliability. Most of the analyses and most of the arguments made by
proponents and opponents of additional capacity to handle large ships
start from assumptions that this conclusion suggests must be suspect.
Assumptions bound the ratios found in benefit/cost analyses.
Decisions with regard to developing additional port capacity, then,
must be made with the recognition that the fundamental reality is an
uncertain future. An uncertain future implies risks:

- If the decision is to do nothing, trade may be lost.
- Or trade may be sustained with existing port capacity but at a
  higher cost.
- If additional port capacity is created, the anticipated trade
  or traffic may fail to develop.
- Or the additional capacity may be insufficient.
- Changes in technology may supersede improvements, and make them
  obsolete.

LONG LEAD TIMES

The nation's dilemma with regard to the port capacity question revolves around the fact that future need is fundamentally uncertain, short-term needs may experience substantial fluctuation from year to year, and developing additional port capacity requires sustained programs carried out over many years. In the case of major federal dredging projects (as pointed out in Chapter 7), the lead time may be up to 22 years. Even assuming that those lead times can be substantially reduced, there is a mismatch between the uncertain and fluctuating character of need and the activities required to develop port capacity. In sum, the nation's decisions with regard to developing additional port capacity must find some accommodation between what will likely be a continuing uncertainty about need and the long lead times required to develop that additional capacity. Stated simply, the nation's choice is: "What should be done in the face of uncertainty?"

Summary of Factors Contributing to Uncertainty

A review of the debate over additional capacity in navigational facilities indicates its complexity. Some of the arguments made by opponents are: (1) The prices charged for ocean transportation have little relationship to costs. Rather, rates vary with surplus and scarcity of vessels relative to cargoes. When vessels are in surplus, unit prices for transport in large versus small ships vary less than the difference between the unit costs. Alternatively, when there is a scarcity of transportation, vessel owners charge whatever the market will bear regardless of cost. (2) The historical movement to larger ships reflects fashion more than economics, but the movement was partly subsidized by nations that (to keep their shipyards busy) moved from subsidizing one generation of ships to subsidizing the next generation of ships. (3) The pattern of developing additional port capacity to handle large ships, again, may reflect fashion as much as compelling economic reasons. (4) The United States is such a major factor in the world economy and in world oceanborne transportation that shipowners will build their ships to ensure that they are able to use U.S. ports. That is, the United States can set the standard for ship size with that standard being existing port capacity and it does not need to develop additional capability. (5) Most of the major competitors with the U.S. for the world's coal market are countries such as South Africa where the government controls the mines, the railroads to the ports, and the ports. Those countries will, as a matter of national policy, ensure that their coal always sells for less than American coal. That is, competitors with the United States for the world coal market will do whatever is necessary to ensure that their coal sells for less than coal from the United States.

   Some of the arguments made by proponents are: (1) Prices charged for transporting commodities do in fact, over the long term, reflect costs. Therefore, the economies-of-scale associated with large ships are reflected in prices. (2) Only if the United States is able to take advantage of low-cost transportation will it be able to maximize its competitiveness in the international economy. (3) Even though other countries may, as a matter of national policy, provide coal for lower prices than the United States, the range of the price differential can make a substantial difference in the U.S. share of the market. If the price differential can be kept small by efficient transport, many countries will pay a marginally higher cost to be assured of the secure, stable supply offered by this country. (4) Faced with a huge trade imbalance of over $100 billion, the United States can pass up no opportunity to increase its exports. (5) Given the long lead times necessary to develop port capacity, the United States has no choice but to move ahead with port development in the face of uncertainty.

The above samples of the opposing arguments made with regard to port capacity and its economic implications are not succeptible to resolution by studies. They reflect differing perceptions of what will occur in the future, of the U.S. role in the world economy, and perhaps most fundamentally they sometimes have an unstated premise. That unstated premise reflects differing views about or whether the federal government should underwrite the costs associated with developing additional port capacity. In sum, much of this debate is driven by differing perceptions of the appropriate role of the federal government with regard to port funding, with the key issue being whether tax dollars should be used for this purpose.

National Defense/Security

Although not a central point in the debate, some have argued that additional port capacity would contribute to the nation's security/defense capability. Like the economic issue, security and defense needs are difficult to ascertain. To the extent that those needs have been defined but are classified for security reasons, this committee has been without the basis for making an informed assessment. The committee did, however, seek information on security-defense needs. Based on those efforts, several observations seem in order. First, security-defense needs exist in three categories: (1) military ships, (2) logistical support for overseas military operations, and (3) access to strategic materials. Responsibility for assuring adequate port capability to meet these three needs rests with different defense and civilian agencies. The Navy assumes responsibility for assuring adequate capacity to handle its own ships. Logistical support for overseas military operations is the responsibility of the Military Traffic Management Command, which has designated the National Defense Ports to be used in case of mobilization. The vessels supporting overseas military operations would operate under the direction of the Military Sealift Command which has responsibility for mobilization of the necessary cargo vessels. Responsibility for assuring adequate capacity to transport strategic materials in times of war or international conflict rests with the U.S. Maritime Administration. The U.S. Army Corps of Engineers has the direct responsibility to ensure the navigability of the nation's ports and waterways.

The committee was unable to find any evidence that these three areas of responsibility were being systematically coordinated or that projected needs for dredging are being communicated to the Corps. Given the changing character of the world's merchant fleet, this is an area which would appear to warrant continuous and careful attention.

For example, in such areas as the ports of Hampton Roads (Norfolk, Newport News, and Portsmouth, Virginia), which might in time of war serve all three of the identified functions, such coordination and planning would appear to be particularly appropriate. In the case of the Navy, the committee found, for example, that the Navy has not specified any need for additional channel depths. Channel depths in that port are 45 ft, yet the Navy has specified that berths for its aircraft carriers should be 50 ft. Before the largest carriers can transit the channels into the Norfolk naval facility, they must unload all their aircraft and pump off most of their fuel. The committee questions whether existing depths are appropriate if rapid access needs to be assured.

In the case of logistical support, the Military Sealift Command expects to use vessels from the U.S.-flag fleet and those of national-flag fleets in NATO. Given the movement toward larger and specialized vessels, a continuing analysis of port navigational facilities would appear necessary.

Finally, as is often noted, the United States is heavily dependent on foreign suppliers for strategic materials. The committee found no evidence that additional port capacity was needed to meet the nation's strategic materials need. Neither did it find that these needs were being coordinated in any meaningful fashion with other defense/security needs and being communicated to the Corps of Engineers.

Future Flexibility

The United States has very limited capabilities to take advantage of any benefits that may be offered by larger ships. Only two of the nation's major ports can handle dry bulk carriers of more than 90,000 DWT and only a limited number of the nation's major container ports can readily handle the latest-generation, high-value cargo vessels. Although evidence is mixed on the rate at which large ships will increase and unclear about what the optimum size of large ships will be, the nation's present capacity to handle these ships is limited. The United States, then, has little flexibility to respond to any developments which emphasize or accelerate the advantages of using large ships.

The nation faces an uncertain future with regard to the quantity of its exports and imports and the mix of cargoes that its ports will need to handle. Its dilemma is that to be able to take advantage of any benefits offered by large ships in the future, it must undertake to develop port capacity now. That is because developing port capacity to handle large ships requires many years. Further, whatever the source of funding, the cost of developing port capacity to handle large ships is high. A decision to develop such capacity, therefore, involves risk. The question faced by the nation is: Should it take that risk?

There is no existing body of data or associated analyses available now or likely to be available in the immediate future which will compel a consensus on this question. The committee found itself in unanimous agreement that faced with this uncertainty, the nation should develop sufficient capability to allow it to be able to respond flexibly to whatever opportunities develop in the future. That is, the United States should move from a position of not being able to accommodate large bulk carriers on the Atlantic or Gulf Coasts and limited capabilities to handle medium-size vessels to one where it has expanded capabilities.

Specifically, it is the committee's conclusion that there should be a capability to handle large bulk carriers on each of the nation's coasts. In present circumstances, that capability could be minimal. The committee can find no justification for the expenditure of federal funds on all of the projects that have been proposed. Should local ports determine it is to their advantage to underwrite the costs for additional development, that is a judgment they should be allowed to exercise.

Alternatively, a limited capacity to handle large bulk carriers on each of the coasts and some expansion of the capacity to handle the medium-size vessels, in the committee's judgment, is in the national interest. In sum, the United States needs additional but limited capability to assure that it will be able to capture the benefits that may develop from being able to handle larger vessels in the future.

REFERENCES

Cargo Systems Research Consultants (1982), Large Dry Bulk Carriers - Employment Prospects in the Eighties (Worchester Park, England, Cargo Systems Research Consultants).

Cushing, C. R. (1984), "The Ships of Tomorrow," Cargo Systems, 11: 32-37.

Energy Information Administration (1983), Port Deepening and User Fees: Impact on U.S. Coal Exports (Washington, D.C.: Government Printing Office).

General Accounting Office (1983), Prospects for Long-Term Steam Coal Exports to European and Pacific Rim Markets (Washington, D.C.: Government Printing Office).

Graves, S. C., M. Horwitch, and E. H. Bowman (1984), "Deep-Draft Dredging of U.S. Coal Ports: A Cost-Benefit Analysis," Policy Sciences, 17: 153-178.

H. P. Drewry (1982), Shipping Statistics and Economics (London: H. P. Drewry (Shipping Consultants), Ltd.).

H. P. Drewry (1981), <u>Governments and Dry Bulk Shipping</u> (London: H. P. Drewry (Shipping Consultants), Ltd.).

MARDATA, Inc. (1984), Survey of World Fleet prepared for Committee on National Dredging Issues, Stamford, Connecticut, Maritime Data Network, Ltd.

Maritime Transport Committee (1984), <u>Maritime Transport 1983</u> (Paris: Organisation for Economic Cooperation and Development).

Office of Technology Assessment (1983), <u>An Assessment of Maritime Trade and Technology</u> (Washington, D.C.: Government Printing Office).

Poten and Partners (1983), "Evaluation of a Port Improvement Project from the Perspective of a Shipowner," Prepared for the Committee on National Dredging Issues.

Schonknecht, R. et al. (1983), <u>Ships and Shipping of Tomorrow</u> (Centreville, Md.: Cornell Maritime Press).

Tozzoli, A. J. and S. Frank (1984), "Federal Channel and Related Development in New York Harbor," Testimony presented to Subcommittees on Energy and Water Development and Senate Committees on Appropriations, U.S. Congress, April 2, 1984.

5
Options for Handling
Large Vessels

Several alternatives have been proposed to accommodate large vessels. This chapter describes five, and compares them by the committee's criteria (economics, navigational safety, environmental effects, national security/defense implications,and contribution to future flexibility). These alternatives are in different stages of development--a few have long been in use, others represent more recent applications of existing technology, and some are speculative.

The five principal options considered by the committee are:

- Construction dredging, that is, underwater excavation of materials to create a wider/deeper channel for larger vessels, can be undertaken at existing ports. It should be understood for the comparative purposes of this chapter that the other options described are not necessarily exclusive of dredging.

- Offshore terminals, designed to accommodate deep-draft vessels, are usually dedicated to a specific commodity. State-of-the-art examples of several types are now in service in many parts of the world.

- The design and construction of new deepwater ports, or offshore industrial islands represent major regional or national commitments. New multicommodity harbors (such as Europoort) and new deep ports for one or two commodities have been built in other countries, and proposals have been advanced in the U.S. for such developments.

- The use of wide-beam ships, with substantial cargo-carrying capacity and drafts compatible with the depth limitations typical of ports in the United States, have been proposed and designed, but not built. Among the problems to be resolved are structural design, maneuverability, and power requirements.

- Lightering of incremental cargo may be used either to lighten incoming ships to drafts compatible with channel depths or to top-off outbound ships. Transfer terminals, enabling rapid and economical transfer from barges to ships, have recently been constructed at New Orleans, and self-unloading barges have been built for top-off services on the East Coast.

Alternative means of <u>cargo transfer</u> accompany several of these
options--dry transfer of crushed or pelletized cargoes by conveyer
belts, tramways, or monorail; pipeline transfer of bulk liquids or
slurried solids--and may, depending on the over-all design and
application, serve as alternatives to dredging. There appear to be no
effective alternatives to inner harbor areas for the transfer of
containers, owing to the large amount of space required (35 acres or
more per containership berth). A trend emerging with operation of the
new large containerships is to designate a major port for calls by
these vessels and to use feeder ships to collect and distribute
containers to and from other ports. While this might have some
implications for dredging, the trend is too recent to evaluate. The
cargo-transfer alternatives described in this chapter have all been
developed for bulk and break-bulk cargoes.

The alternatives described and compared in succeeding sections need
to be evaluated in the specific contexts of their proposed applications
for capital costs; compatibility with existing infrastructure, and with
existing land and sea transport; operational costs; environmental
considerations; political considerations; and safety. No general
characterization can be made of the dominating considerations or unique
circumstances that would guide the choices to be made in a particular
location.

CONSTRUCTION DREDGING

As dredging is the principal subject of this report, various aspects
are described in detail in other chapters, and only a brief description
is given in this section. In dredging, materials below the water
surface are excavated and transported to a designated site for
placement. Navigational channels, maneuvering areas, anchorages, and
berths have been created and maintained by dredging for many hundreds
of years. "Construction dredging" refers to excavation of virgin
materials either to create or to improve (by deepening, widening, or
lengthening) these navigational facilities.

In the context of dredging to accommodate large vessels, it bears
mention that while the placement of dredged material is often
incidental to the objectives of dredging, it is sometimes the primary
objective--creating landfill, for example, to expand terminal
facilities. In the proposals of some ports for construction dredging,
enlarging the navigational facilities and creating landfill for new or
expanded terminals are equally important in accommodating large vessels.

OFFSHORE TERMINALS

Offshore terminals have long been recognized as a potentially optimum
solution for bulk cargoes. They are usually designed for one
commodity, restricted to use by large vessels, and equipped with
modern, efficient handling and transfer facilities.

42

An advantage of these terminals is that they are located offshore, at a distance from concentrations of people and facilities, and are usually in open waters. Thus, a disaster such as fire or explosion would be remote from areas of extreme vulnerability, and accidental oil spills and airborne contaminants would be more likely to be dispersed. These advantages depend to some extent on remoteness, which may add to the cost of an offshore terminal, and in the case of dispersal, on winds and waves that may not be favorable. Generally, however, offshore terminals can greatly reduce the risks of certain cargoes to concentrated human populations and to restricted environments, such as estuarine marshes. Because the characteristics of the commodity are known, specialized safety and pollution-prevention equipment can be installed.

In a discussion of the advantages and disadvantages of offshore terminals, Soros (1983) emphasizes the protection of coastal resources as the primary advantage, and among the problems, lists the following for marine operations:

- The tug-ship operation in the open sea
- The docking approach, especially with adverse winds, waves, and currents
- Attachment of mooring lines to buoys, dolphins, and pier structures, and detachment, especially from mooring buoys
- Decision about when to leave berth owing to worsening weather

The principal constraint acting against construction of offshore terminals is their high capital costs. The Japanese contribution to an international study (Permanent International Association of Navigation Congresses, 1977) stated opposition to uncoordinated planning of offshore oil terminals, and support of joint operation "to ensure the efficient operation of berths, facility of construction, and safety to maritime traffic." Many offshore terminals (for example, the Drift River Terminal, Cook Inlet, Alaska) are operated by consortia, and others are operated as public utilities.

Offshore terminals have occupancy rates ranging from 65 percent to 90 percent, depending on exposure and design. Occupancy is important in determining the economic promise of such a facility, since unusable time incurs extra costs for the terminal and the vessel.

The principal types of offshore terminals are described briefly in succeeding sections.

Fixed Offshore Structures

Fixed offshore structures, against which vessels berth for cargo transfer, are supported by piles or caissons. Delivery to (from) shore is usually by pipeline or conveyor. Three examples, among many worldwide, are:

- The offshore oil terminal at Tomokomai, Hokkaido, Japan, located 5 miles offshore in the Pacific Ocean, where prevailing winds

blow seaward, accommodates very large crude carriers up to
350,000 DWT. A pneumatically controlled oil boom is raised from
the seafloor as the ship docks. Vacuum suctions prevent oil
spillage when the transfer lines are disconnected.
- The coal export terminals two miles offshore Hay Point, MacKay,
  Queensland, Australia. Coal is transferred on covered conveyor
  belts. Stockpiles on shore are sprinkled to reduce dust.
- The offshore iron ore export terminal at Port Latta, Tasmania,
  Australia, where pelletized iron ore is conveyed two miles out
  to a terminal in the open sea for loading into large ore
  carriers.

A fixed terminal was proposed for liquefied natural gas (LNG) in
waters offshore southern California to minimize the risks to coastal
populations of an explosion or ignited cloud of gas. The proposal
called for regassification of the LNG and transportation ashore by
submarine pipeline.

Spread-Type Offshore Moorings

An example of a spread-type offshore mooring is located offshore El
Segundo, California. Incoming tankers moor to a system of buoys. The
hose to the submarine pipeline is lifted from the seafloor and
connected for discharge. This system requires tug assistance in both
mooring and cutting loose. The time for connecting and disconnecting
is correspondingly longer, and the use of the berth is restricted to
calm and moderate seas, with winds from westerly directions.

Single-Point Moorings

The ship moors to a single-point buoy or articulated arm, and the hose
is brought aboard for connection. The ship swings or "weathervanes"
about the buoy. This system is widely used throughout the world for
oil shipment and (to a lesser extent) for oil imports. Examples of the
latter are at St. John, New Brunswick, where the 30-ft tides would make
any fixed terminal very expensive. The single-point mooring has also
been used for iron-ore slurry transfer off New Zealand, described by
Wasp (1983):

Placed in operation in 1971, the Waipipi Iron Sands project consists
of a relatively short land pipeline and a ship-loading system. A
1.8 mile undersea line carries the slurry to an offshore mooring
buoy, where special Marconaflo tankers load the slurry and then
de-water it on board. After the transport water is removed from the
cargo and piped ashore, the tanker sails to its destination with the
iron ore.

44

Floating Terminals

Floating storage and transfer facilities include the Arco Seki Ardjuna terminal for liquefied petroleum gas (LPG) in the Java Sea, and numerous floating offshore oil storage vessels in Indonesia, West Africa, the Persian Gulf, and elsewhere. Vessels are usually moored to an articulated arm or single-point mooring, so as to weathervane. For transfer operations, vessels come alongside or astern. Similar floating terminals have been proposed for LNG service.

Offshore Storage Caisson Terminals

This category includes the Dubai steel tanks placed underwater for oil storage, and the many large concrete caisson structures in the North Sea, which store oil for transfer through an associated articulated mooring buoy. Direct offloading concepts have been developed but not so far used because of concerns about tanker override.

Offshore Terminal Complexes

The outstanding example of an offshore terminal is the Louisiana Offshore Oil Port (LOOP). A complex of single-point moorings is connected by submarine pipelines to a control and pumping terminal structure for transfer by submarine pipeline to onshore and inland refineries. A similar project has been proposed for Freeport, Texas, to handle 500,000 barrels of oil per day. These "superport" projects have been hard hit by the decline in demand for imported oil (in the United States) and by the withdrawal of federal financial support. This has dimmed the prospect for additional building and expansion. Worldwide, most offshore terminals are located in countries other than the United States, yet the majority were designed by engineers and in many cases, constructed by contractors from the United States. This is in large part because such terminals are designed for one commodity, and sometimes for dedicated vessels, and are financed on the basis of export or import of that single commodity.

NEW PORTS; INDUSTRIAL ISLANDS

Construction of entirely new ports with deep-water capabilities is one means of accommodating large modern vessels. Historically, new ports or industrial islands have been planned as part of economic development projects—the import of raw materials, for example, being coordinated with onshore processing or manufacturing. New ports have been developed in Japan in the last several years for petroleum, petrochemicals, and alumina-bauxite. Examples of such developments are the East Harbor at Tomokomai, Hokkaido, Japan, and Sines Harbor in Portugal.

A much larger industrial-island development was studied for the Netherlands, with the conclusion that the high capital costs could not be justified. Industrial islands to handle oil imports have been proposed for the New York Bight Apex and Port Angeles, Washington. Neither has attracted investor interest with the recent decline in imported oil. Los Angeles, California, has conducted preliminary studies of an offshore "Energy Island" that has the principal aim of moving the transfer of hazardous and potentially polluting cargoes away from the densely populated Inner Harbor area. The Port of Los Angeles has long recognized a "hazardous footprint" in the proximity of terminals and tanks to fish canneries and public harborfront developments.

In the United States, development of a new port in the lower Delaware Bay was the subject of a recent conference (Firstport, 1984, proceedings in preparation). The obstacles that would have to be overcome by such an ambitious plan far exceed those of dredging existing ports in the United States (and indeed, dredging and land reclamation would be significant in creating this new port).

WIDER-BEAM CARRIERS WITH RESTRICTED DRAFT

Increasing any dimension of a vessel can significantly augment cargo-carrying capacity. One alternative to deeper ports, therefore, is large vessels of shallower draft and wider beam. Hydronautics, Inc. (1982), Roseman (1979), Roseman et al. (1974) and Roseman and Barr (1984) describe the development and analysis of designs for dry bulk carriers of 60,000 DWT to 200,000 DWT, having restricted drafts of 35 to 55 ft.

Relatively wide-beam vessels have been built and operated, but most of those 100,000 DWT or more are not restricted-draft designs, and exceed the water depths of almost all U.S. ports, fully loaded.* An exception is the Amoco Trinidad-class tanker, 150,000 DWT and 50 ft draft, which can call on West Coast tanker ports (Valdez, Los Angeles, Long Beach) fully loaded, and on Gulf Coast ports somewhat light-loaded. (Ono et al., 1985). The existing vessels and the restricted-draft design series follow the dimensional ratios required by classification societies and considered reasonable limits by naval architects.

Departures from these limits have been suggested: Mitsubishi (1974) proposed an ultra-shallow-draft vessel, a 105,000 DWT bulk carrier that would have a 64 m (210 ft) beam and 10 m (33 ft) draft. There are some concerns with radical designs; for example, as the structural design of the hull will no longer follow the semi-empirical rules of ship

---

*British Steel, 173,000 DWT, 58 ft draft; Shinho Maru, 208,952 DWT, 60 ft draft; BHP ore carrier, 222,000 DWT, 60 ft draft.

classification societies, design criteria for torsion due to
quartering waves would be required.*

Wide-beam vessels are sensitive to underkeel clearance, and can be
more difficult to maneuver in restricted waters (Landsburg et al.,
1983; Roseman et al., 1974); however, this problem is overcome on
existing (single-screw) vessels by much larger rudders. The
Mitsubishi ultra-shallow-draft design is dual-propeller, dual-rudder,
reminiscent of an articulated tug-barge combination.

Very wide-beam vessels (140,000 DWT or more) of greatly restricted
draft could require additional channel widths, and may exceed the
reach of existing loading equipment for dry bulk commodities in U.S.
ports.

The advantage of wide-beam, restricted-draft vessels is that given
a specific draft limitation, they are more economical than smaller
vessels with conventional proportions. However, the advantage of
larger, restricted-draft configurations tends to diminish as they
approach extreme proportions. Given a specific deadweight
requirement, the greatest transport economy (subject to voyage
constraints and practical design limitations) is obtained with
conventional deep-draft vessels. For cargo-carrying capacities above
100,000 DWT, foreign shipowners have so far been unwilling to accept
the draft limitations of U.S. water depths as a design criterion.

LIGHTERING/TOPPING-OFF

Lightering--unloading part of a vessel's cargo to allow it to proceed
at lesser draft--has been practiced for hundreds, perhaps even
thousands, of years. Until recently, lightering operations have
involved low rates of cargo transfer. New self-loading and -unloading
barges and bulk transfer facilities have transformed this ancient
practice, particularly for the reverse operation of loading vessels in
deeper water with the extra cargo that could not otherwise be loaded.
Midstream transfer facilities to handle barge-to-ship transfer at high
cargo rates are now in operation in New Orleans for most bulk
commodities. A topping-off service to load coal into large bulk
carriers unable to load fully in the ports of Hampton Roads, Virginia,
has been developed for the lower Delaware Bay, in an anchorage area
used to lighter oil from larger to smaller tankers (Dowd, 1983). The
advantages of lightering/topping-off services such as these are:

- Potential for use by several ports and many shippers
- Low capital investment
- Rapid and progressive implementation
- Costs borne directly by users
- Flexibility to use topping-off vessels in other functions

---

*Hull forms similar to that of the ultra-shallow-draft design have
been built for heavy-lift roll-on/roll-off vessels, but not for
tankers or bulk carriers.

The disadvantages, depending on particular circumstances, are:

- Higher risk of fire and spills, owing to higher exposures
- More vessel time required to load cargoes
- If carried out in the open ocean, dependence on weather

CARGO TRANSFER TECHNOLOGIES

An especially attractive means of cargo transfer is emerging--the use of slurries to transport coal and iron ore--and it may enhance the use of offshore terminals. Other materials, such as copper ore, can also be slurried, but the volumes do not usually justify the use of an offshore terminal.

The most commonly used slurry medium is water, but oil, methanol, or liquid carbon dioxide might be used. As noted, iron sands in New Zealand are slurried and loaded in dedicated vessels. Slurry transportation of coal has been demonstrated (Wasp, 1983), and can readily be applied to the loading of vessels (Bertram, 1982).

Roseman (1979) states that slurry carriers for pelletized ores and other dry bulk commodities in converted or new special bulk carriers can be considered state-of-the-art technology, owing to the following advantages:

- Reduced cargo handling time between ship and terminal
- Elimination of airborne pollution from mechanical handling of coal
- Ability to operate from offshore terminals employing single-point moorings, permitting the use of large deep-draft vessels, with corresponding economies of scale
- System compatibility with proposed coal slurry pipelines

Among the several proposals, coal slurry would most likely be pumped in 50 percent concentration (by weight) from shore storage to special slurry tankers via submarine pipelines, and dewatered to about 75 percent coal, 25 percent water. Closed-circuit pipelines could be used to eliminate discharge to the sea.

Harris (1983) discusses slurry loading of vessels as an alternative for long-term improvements to coal-exporting facilities in Australia. Among his principal recommendations is: "Loading of coal by pumping slurry or capsules to offshore berths at single buoy moorings." This is an adaptation of the offshore terminals developed for crude oil and petroleum products. The advantages of such a system would be:

- The pipeline would cost less than a jetty equipped with a conveyor
- The loading point can be farther offshore and accept larger ships
- The loading equipment would be less expensive

- The time to build the facility would be reduced
- As with other offshore facilities, adverse effects might possibly be lessened for sensitive environments

While slurry transfer of iron ore, copper ore, and coal has been well established for long-distance transfer over land, its use in loading ships is relatively new. However, it has been thoroughly engineered, and shows promise as a practical and economical mode for the future.

Tramways and Monorails

The use of tramways and monorails for transfer of dry cargo from shore to ship, or ship to shore, has been studied periodically by the U.S. Army and Navy, primarily as a means of unloading military cargoes onto undeveloped coasts. The commercial application of these techniques for loading bulk carriers has been studied and rejected; for example, to load iron ore in Goa, India. These systems have inherent limitations: slow loading rates, problems with excavation from the holds, and potentially high maintenance costs.

Because of their limited capacity, tramways are limited in use to high-density, high-value cargoes such as copper ore. It is uncertain whether sufficient commercial or military interest exists to carry forward the engineering development of tramways or monorails.

COMPARISON BY THE COMMITTEE'S CRITERIA

In this section, the previously described alternatives are compared by the committee's criteria--economics, navigational safety, environmental effects, national security and defense, and contribution to future flexibility. This comparison is necessarily general: a detailed comparison could only be made on the basis of well-developed plans for each alternative.

Economics

Of the alternatives considered in this chapter, the construction costs of a new deepwater port would appear to be the greatest and those of topping-off services the least. Estimates have not been developed for the only recent new-port proposal in the United States (in the lower Delaware Bay). A five-year, $10 million study of the Delaware Bay proposal has been suggested (Gaither, 1983). Owing to the number of existing ports and their competition with one another for cargoes, it has been remarked (Firstport, 1984) that several ports in a region would have to participate as investors in the development of a new port, and continue to support it with feeder and transshipping services. A group of ports in a region might not be able or willing to make such a large investment: it is possible that some combination of federal, state, and local investment guarantees would be necessary, as well as formation of a public-private, or very large private consortium to finance and manage such a project. Prospects for such developments in the United States appear questionable.

As indicated in Chapter 4, the construction costs for dredging the five ports with proposals for dredging to 50 ft or 55 ft depths range between $278 million and $440 million. The range of estimated construction costs for dredging each of the five ports proposing to accommodate the latest-generation containerships is between $3 million and $80 million. For containerships, there appear to be no effective alternatives (other than new deepwater ports) to dredging existing ports.

For liquid- or dry-bulk commodities, the economic advantage of deepening existing ports and harbors in comparison to the non-port alternatives is that loading and unloading is assured if the ship can be brought to port. Offshore alternatives are all weather-sensitive to greater or lesser degree, and this can affect ship schedules. Delays may not be as large an economic factor as other factors, however. Since the alternatives tend to address specific needs, their economics tend to be specific to location, commodity, timing, intended throughput, and actual volumes handled.

The offshore terminals already installed in waters of the United States are all oil terminals, and these are not in every case competitive with transshipment from larger into smaller tankers for unloading in existing ports. This may be due to timing or to a combination of factors that different market conditions would reverse.

Proposals for coal slurry pipelines to carry coal from mines to deepwater terminals have estimated costs of $140 million to $750 million (Bertram, 1982). Coal slurry transportation and loading systems would depend for their economic competitiveness in part on achieving lower inland transportation costs. The costs associated with the terminal end of such systems are almost certain to be higher than those of already existing terminals, and could only be competitive in loading large bulk carriers. To pay for themselves, offshore coal terminals would need to handle substantial volumes.

A wide-beam vessel of 120,000 DWT capacity and 38 ft draft has been estimated to cost $120 million to build in the United States (1980 dollars). Estimates have not yet been developed for larger wide-beam vessels with very shallow drafts (Bertram, 1982). Large vessels of extreme proportions may have hidden system costs, and may not be competitive with already existing large bulk carriers. Except in special cases, given the worldwide surplus of very large carriers of oil, coal, and ore, and hence, the low cost of using already existing large carriers, constructing a fleet of vessels to such specifications appears unlikely except for "captive" runs. One of the principal reasons for design studies of these vessels in the United States was to give the U.S.-flag merchant marine large bulk-carrying capacity. This motivation has been clouded by the uncertain economic return that could be expected from such vessels in competition with the existing heavily overtonnaged world bulk-carrier fleet, the probable need to reserve a percentage of U.S. coal exports for the vessels, and the government's announced intention to eliminate subsidies for the construction and operation of U.S.-flag vessels.

Of all the alternatives considered, topping-off has the shortest lead times and lowest capital investment. Investments have already been made in three technologies (floating terminals, self-unloading ships, and self-unloading barges), and this existing capability enables short-term response to the need to load large-volume bulk carriers. To gain the needed return on the investment depends on the willingness of shippers and shipowners to pay the additional cost, and spend the additional time for topping-off operations. Thus, the charge per ton (averaged over total tons carried) cannot exceed the transportation cost-savings per ton of using larger vessels, and for the past three years, this has represented a relatively small difference.

The economically most attractive alternatives for accommodating large vessels appear to be construction dredging of existing ports and lightering/topping-off.

Dredging of existing multicommodity ports is attractive for the following reasons:

- Economies of scale associated with large vessels are provided for all commodities and cargoes which could benefit
- Existing multicommodity, multipurpose ports offer economic protection against the volatile fluctuations of trade in single commodities
- Existing ports represent already sizable investments in terminals and other port facilities and services (described in Chapter 4), as well as established infrastructures of inland transportation. Among the existing services of ports are worldwide sales organizations (allowing them to pursue vigorously whatever cargoes are available)

There are, of course, economic risks associated with dredging: the capacity created in anticipation of demand may exceed actual demand or fail to serve it, and the emergence of new technologies could make improvements obsolete.

Navigational Safety

As with economic issues, the navigational advantages and disadvantages of various alternatives for accommodating large vessels cannot be assessed in detail without well-developed plans. Any engineered system represents a set of compromises between several goals; for example, between project cost and safety, and forces and features of a particular environment. Following is a summary of experience with the trade-offs which must be made with each of the alternatives.

New Construction Dredging

Dredging existing ports offers the opportunity to enhance the safety
margins of vessel operations in approach channels and within the
sheltered waters of a port. New construction dredging also offers the
opportunity to accommodate even larger vessels and more traffic in
these navigational facilities with smaller margins of safety. Thus,
the contribution to navigational safety of new construction dredging
in existing ports depends on adequate design, maintenance, and
operational practices.

Offshore Terminals

Offshore terminals are located in deep water, as opposed to the
protected waters of coastal ports and harbors, but their relatively
greater exposure has not resulted in higher rates of casualties than
similar port and harbor operations. Docking and undocking may be
simpler than the equivalent operation at port terminals, and offshore
terminals may reduce port vessel traffic.
    A particular problem for offshore terminals (if tug assistance is
required) is the availability and capability of oceangoing tugs:
those that are used to assist harbor maneuvers are not designed for
open ocean conditions, and the tugs designed for oceangoing functions
are not designed for the maneuverability of docking and undocking
operations.

New Deepwater Ports

The contributions of a new deepwater port to navigational safety could
be substantial, but this would depend on the relative importance of
this criterion as a design goal. The opportunity to enhance
navigational safety by providing ideal channel layouts and approaches
may be considerable. Any comparison would have to take into account
the specific circumstances and characteristics of existing ports in
the region that might be improved versus those of the new port.

Wide-Beam Vessels

The preliminary tests of extremely wide-beam, shallow-draft vessels
for inherent controllability by computer simulation (Eda, 1983,
Aranow, 1983) indicate that they are unresponsive. The problems
exhibited in simulated maneuvers can probably be solved by very large
rudders, rudders of different design, and other adjustments, but these
could also affect the vessels' economics in greater power
requirements. Large wide-beam vessels with drafts of 50 ft to 60 ft
have been built and operated successfully. Depending on their
dimensions and trade, such vessels could require additional dredging
in many U.S. ports.

A large wide-beam vessel of shallower draft could require channel widening for navigational safety. An additional navigational concern for wide-beam vessels is narrow bridge openings. These already present a navigational hazard in many U.S. channels (Marine Board, 1983).

Topping-Off

Topping-off or lightering operations carried out in semi-protected or unprotected waters (where sufficient water depth is available) imply some dependence on weather. For midstream loading, winds acting on the unloaded or lightly loaded vessel are perhaps the factor of most concern in operations. Generally, the navigational safety of lightering, topping-off, and midstream transfer are about the same as for offshore terminals, with the obvious difference of involving two vessels, and for oil transshipments, have been carried out for many years without major casualties.

Environmental Issues

While the potential environmental effects of some alternatives for handling large vessels are associated with construction or the disposal of dredged material, others are principally associated with the vessels and their cargoes. The committee has not assessed or compared these latter risks in detail.

The potential environmental effects of any alternative are highly site-specific, and adverse effects may be averted or mitigated by conscientious planning, siting, engineering, and operations. Some general observations about the potential environmental effects of the alternatives are offered here.

New Construction Dredging

The environmental implications of new construction dredging are discussed in Chapter 9. These vary with the specific characteristics of the project, the characteristics of the physical and biological environment(s) of the project and disposal sites, and other factors, and can only be known in site-specific studies. Some of the general points made in Chapter 9 deserve mention for comparative purposes: (1) potentially adverse environmental effects can be caused by dredging and the disposal of dredged material; (2) adequate planning, design, action, and follow-up activities give reasonable assurance of minimizing and managing the environmental consequences of dredging and disposal of dredged material (that is, an adequate base of scientific and technical knowledge exists to guide decisions and action); (3) dredging may have some environmental advantages--removal of contaminated materials (if properly managed), beach replenishment, and wetlands rehabilitation.

Offshore Terminals

As indicated in preceding sections, there are environmental advantages to locating certain terminals offshore, and those of the United States are sufficiently far from shore to ensure maximum protection of coastal resources and concentrations of population from catastrophic or operational pollution and accidents. For coal terminals offshore (none now exists in the U.S.), questions of environmental implications might be raised about the fluid medium in the slurry, and its ultimate fate and effects.

New Deepwater Ports

It is likely that the greatest change to the local environment from the creation of a new port would occur with shoreside developments, particularly as the sites proposed for new-port development have little existing landside infrastructure. Dredging would also be required (both construction and maintenance) in these locations, even though existing water depths are greater than the natural depths of existing ports, to create berths and other facilities, and to make depths uniform. How much dredging would be required (and the environmental effects) depends on the design and layout of the port, and on its site-specific characteristics.

A detailed risk and consequences assessment would be necessary to determine the level and severity of hazards posed to the environment and surrounding populations of vessel casualties.

Wide-Beam Vessels

Environmental implications of wider-beam ships would appear to be comparable to those of full-form tankers, with one major exception, and that exception is that channels may have to be widened rather than deepened, particularly if two-way traffic is desired. As with any dredging project, the potential exists for adverse environmental effects.

Topping-Off

Concerns have been expressed about the environmental effects of topping off large bulk carriers with coal in the lower Delaware Bay (Biggs et al., 1984). These concerns center on the fates and effects of coal lost to the air and water in the transfer operation. Environmental concerns have not been raised for topping off in the Gulf of Mexico, or for midstream transfer in the Mississippi River, where the operations are viewed as comparable to coal loading at a port terminal (Chatagnier, 1983). There are no ports in the lower Delaware Bay: the anchorage proposed for coal topping-off has been used for oil transshipment and is regulated by the U.S. Coast Guard.

The Coast Guard conducted an environmental assessment in reviewing the permit application (the permit was granted), but authority over the environmental quality of Delaware's coastal waters was claimed by the state Department of Natural Resources and Environmental Control and Delaware law has been interpreted to prohibit the proposed topping off activities.

National Security and Defense

Much of the nation's contingency planning for national security and defense involves oceanborne transportation. The Joint Chiefs of Staff estimate that for any major overseas deployment, 95 percent of all dry cargo and 90 percent of all petroleum will move by sealift. The armed forces appear to have given little attention to the security- and defense-related aspects of port dredging in the United States and the proposed alternatives (General Accounting Office, 1983). Review of these aspects is therefore somewhat speculative.

As noted in Chapter 4, some combatant vessels have greater depth requirements than are now provided by the navigational channels they use or propose to use. While operational flexibility (such as waiting for high water, light-loading and one-way traffic) may be adequate to ensure transit in many of these situations, it may not be adequate in all situations. The same considerations may apply to the noncombatant vessels used. If the capacity to accommodate vessels with greater depth requirements is needed by some combination of combatant vessels and those to be used for mobilization or the transport of strategic materials, then new construction dredging of existing ports offers the maximum contribution to national security and defense (the existence of ports that are able to respond to all three needs broadens the nation's capability to support these activities).

The possible contributions of offshore terminals is somewhat equivocal: they may present a valuable option or an additional vulnerability. They may also have no potential contribution, positive or negative, depending on the commodities handled. New deepwater ports, on the other hand, could offer the opportunity to include defense facilities or features that might be difficult to achieve at existing ports, but this is a contingent opportunity that has not been addressed.

There seem to be few implications for national security or defense from the introduction of large wide-beam vessels. Alternatively, the flexibility of topping-off and lightering services could prove important in moving strategic materials or petroleum.

Future Flexibility

As implied in preceding sections, new construction dredging of existing multicommodity ports gives the nation the greatest future flexibility among the options. Assuming that offshore terminals are for single commodities, they offer little additional flexibility for the United States, and it might be argued that they increase the nation's inflexibility. On the other hand, they are a proved technology that can be provided to the pipeline infrastructure without demands for space on the ports' waterfront, and this enhances flexibility.

A new deepwater port could clearly add to the nation's future flexibility, if it were planned as a multipurpose port handling a mix of cargoes. Alternatively, if the deepwater port were primarily a single-commodity port, it would offer less future flexibility than the deepening of an existing multipurpose port. If the investment required that the new port have a regional monopoly on port services, however, future flexibility would clearly be reduced.

The contribution of wide-beam ships to future flexibility appears minimal, since they do not seem to be competitive with deep-draft vessels.

The lightering/topping-off option on the other hand, offers the nation a short-term, low-cost response to handling large-volume bulk carriers. It appears attractive where volumes of dry-bulk commodities are sufficient to repay the investment. It is not likely to be offered for certain commodities in particular circumstances, nor can it substitute for additional channel depths to accommodate containerships. This option contributes to present as well as future capacity to accommodate large-volume vessels, but cannot be expected to meet all the needs projected.

CONCLUSIONS

Of the five options for increasing the nation's capacity to handle large vessels, measured against the criteria of economics, navigational safety, environmental implications, national security/defense needs, and future flexibility, two of the options stand out as being the most attractive. That is, assuming the conclusion drawn in Chapter 4 is correct that a prudent society can ill afford to move into the future without the capacity to handle large ships, then lightering/topping off and dredging existing ports are clearly the most attractive two options.

This necessarily general judgment does not exclude any of the other options, which may be attractive for some particular application now or for very different circumstances in the future.

Lightering/topping-off is a developed and available technology that is sufficiently flexible to meet short-term contingencies and serve developing needs. Its short lead times allow the market to measure its attractiveness.

Alternatively, new construction dredging of existing ports cannot respond to short-term market changes. The lead times for new construction dredging are long, and the short-term future uncertain. The reason for developing at least a limited program of new construction dredging in this country at existing ports is that dredging offers the most secure response to an uncertain future. That more secure future results from the fact that construction dredging puts the United States in a position to take advantage of the changes in maritime transportation that have already occurred, those that can be projected in the near term, and those that may occur.

Two optional categories of dredging to handle larger vessels can be distinguished. One is dredging to accommodate vessels requiring 40 to 45 ft of water depth. The most frequently cited need is that of the latest-generation containership, but there are many vessels in the world fleet in this medium-size category, and the container ports that handle a range of commodities could benefit from being able to accommodate more and larger vessels of all types. The evidence presented by several of the major ports in the United States and that reviewed by the committee suggest that these are needs that exist now. Given the heavy existing investment in cargo-handling facilities in these ports, the well-established inland transportation systems serving them, and the expectation that the quantity of cargoes carried in medium-size vessels will increase, there is compelling reason to assure that construction in this medium-depth range can be carried out by the ports that can justify it.

The second category includes dredging to accommodate larger bulk vessels requiring depths of 50 ft or 55 ft (or more). The committee concludes that the prudent choice, given an unpredictable future, is to ensure the nation has future flexibility. This conclusion does not entail dredging all the proposed deep-water bulk-commodity projects that have been put forward, nor does it suggest the number or order of ports to be dredged to depths of 50 or 55 ft or more. The future flexibility criterion does suggest that the nation should have a minimum of deep-draft capability on each of the coasts.

Criteria Applicable to Selection of Ports for Deep Construction Dredging

If, in the face of uncertainty, prudence suggests that additional dredging of existing ports to achieve deep facilities is necessary, what criteria might be used in establishing priorities for construction dredging? Four criteria seem compelling. First, major emphasis should be given to the ports with multicommodity capabilities. Ideally then, selected ports should be capable of handling all types of high-value cargo vessels ranging from containerships through roll-on/roll-off ships, to break-bulk vessels. Similarly they should include ports with facilities capable of handling coal, hard minerals, grain, and oil.

Second, consideration should be given to the adequacy of the inland transportation systems serving the ports. Two factors should be considered here. First, the most attractive ports would be those that serve a range of economic activities--manufacturing, agriculture, mining--and the greatest numbers, in terms of population or markets. Consideration should be given to the availability of alternative inland transportation systems. The ideal would be a port which is served by highways, multiple rail lines, and inland waterways. Where there are alternative inland transportation options, competition offers the best prospect of keeping inland transport prices low. One concern frequently expressed is that a deep-water port served by a single inland transportation system might see that system increase transportation costs to the point of cancelling the economic advantages of using large ships.

The third criterion that should be considered is the comparative cost of construction dredging and the additional maintenance dredging costs that will have to be met annually (or at whatever the maintenance dredging interval).

An important consideration in all port dredging decisions is the potential effect on the environment. Thus, the fourth criterion for selection among candidate ports for deep-draft dredging is minimizing the potentially adverse environmental consequences.

REFERENCES

Aranow, P. I. (1983), "Maneuvering Response Supplemental Experiments (Collier and Containerships)," CAORF Technical Report 42-8218-01, National Maritime Research Center, Kings Point, N.Y.

Betram, K. M. (1982), "Alternatives to Deep-Draft Port Dredging for U.S. Coal Export Development: A Preliminary Assessment," Report No. ANL/EES-TM-183, Argonne National Laboratory, Argonne, Illinois.

Biggs et al. (1984), "Coal Transfer: Can an Environmentally Safe Coal Transfer Operation Be Undertaken in The Lower Delaware Bay?" Report No. DEL-SG-01-84, University of Delaware Sea Grant Program, Lewes, Delaware.

Chatagnier, G. (1983), "Midstream Mooring Facilities," Ports '83, K. Wong, ed. (New York: American Society of Civil Engineers), pp. 402-414.

Dowd, J. P. (1983), Presentation to National Coal Association Conference, New Orleans, September 19, 1983.

Eda, H. (1983), "Shiphandling Simulation Study During Preliminary Ship Design," Proceedings, Fifth Annual CAORF Symposium, Kings Point, N.Y., May 12-13, 1983.

Gaither, W. S. (1981), "A National Deepwater Port in Delaware," Marine Policy Reports, 4: 1-4 (University of Delaware, College of Marine Studies).

General Accounting Office (1983), Observations Concerning Plans and Programs To Assure the Continuity of Vital Wartime Movements Through United States Ports (Washington, D.C.: Government Printing Office).

Harris, A. J. (1983), "Marine Works for Bulk Loading," The Warren Center, University of Sydney, Sydney, Australia.

Hydronautics, Inc. (1982), Advanced Technology U.S. Flag Bulk Carriers, Report prepared for U.S. Maritime Administration (Laurel, Md.: Hydronautics, Inc.).

Landsburg, A. C. et al. (1983), "Design and Verification for Adequate Ship Maneuverability," Paper presented at Annual Meeting, Society of Naval Architects and Marine Engineers, New York, November 9-12, 1983.

Marine Board, National Research Council (1983), <u>Ship Collisions with</u>
<u>Bridges: The Nature of the Accidents, Their Prevention and</u>
<u>Mitigation</u> (Washington, D.C.: National Academy Press).

Mitsubishi Heavy Industries, Ltd. (1974), "Ultra Shallow Draft Vessel."

Ono, M. et at. (1985), "The Design of Tankers for Restricted Draft
Service," Paper presented to STAR Symposium, Society of Naval
Architects and Marine Engineers, Norfolk, Va., May 21-24, 1985.

Permanent International Assembly of Navigation Congresses (1977),
"Report of the Japanese National Section," <u>Proceedings of the 25th</u>
<u>International Navigation Congress</u>, Leningrad, U.S.S.R.

Roseman, D. (1979), "Relative Costs of Alternative Modes of Ocean
Transport of Coal," Presentation to the Third International
Symposium, Transport and Handling of Minerals, Vancouver, Canada,
October 21-24, 1979.

Roseman, D. P. and R. A. Barr (1984), "Restricted Draft Geometry--
An Alternative to Dredging," <u>Oceans '83</u> (Washington, D.C.: Marine
Technology Society and Institute of Electrical and Electronics
Engineers, Inc.).

Roseman, D. P. et al. (1974), "Characteristics of Bulk Product
Carriers for Restricted Draft Service," Presentation to Annual
Meeting, Society of Naval Architects and Marine Engineers, New
York, November 14-16, 1974.

Soros Associates (1983), "Report on Offshore Materials Handling
Terminals," New York.

Sugin, L. (1972), "Alternatives in Port Terminal Layout--Dredging vs.
Offshore Terminal," Society of Mining Engineers of AIME, Preprint
72-B-68.

Wasp, E. J. (1983), "Slurry Pipelines," <u>Scientific American</u>, <u>249:</u>
48-55.

6
Funding Issues

The port dredging stalemate results from a complex and interacting set of factors. Three barriers to new construction dredging or increased maintenance dredging are regularly identified: (1) lack of national funding, (2) institutional problems, and (3) environmental problems. This chapter addresses the first barrier.

As discussed in Chapter 3, since 1824 the U.S. Army Corps of Engineers has had primary responsibility for both new construction and maintenance dredging of ports in the United States. Until 1970, there was a consensus that dredging would be paid for from general U.S. Treasury revenues. Beginning in the early 1970s, that consensus on funding began to unravel.

Prior to the 1970s, struggles over port dredging occurred primarily in the context of the annual congressional appropriations process. The primary issues that had to be resolved each year concerned the level of appropriations and which ports should receive construction funding. Since 1970, the struggle over dredging has experienced a fundamental change. At issue now is whether additional new construction dredging is needed, and if it is, what the source of funding should be.

No agreement exists about why the traditional consensus on funding came to an end. Four factors, however, are repeatedly identified as contributing to the erosion of the legislative consensus that funding should come from general revenues:

- the federal budget deficit
- the high cost of new construction dredging (and possibly increased maintenance dredging)
- problems of initiation after a long stalemate
- changing social values and attitudes

These four factors are discussed in the succeeding section.

FACTORS CONTRIBUTING TO THE FUNDING STALEMATE

The most frequently identified factor in the funding stalemate is the growing size of the federal budget deficit. Starting in the 1970s and

59

continuing into the 1980s, concern with deficits was driven, in part, by the unpredictable performance of the national economy. With declining confidence in the economy's ability to sustain predictable rates of growth, it was no longer possible to assume ever-growing tax revenues. In parallel, so-called nondiscretionary activities in the federal budget manifested a pattern of seemingly uncontrolled growth. One consequence of these two patterns was a growing concern with a future of large and rising budget deficits.

In this environment, categories of federal expenditures that were perceived as being discretionary received increasing congressional attention. In the eyes of many, water-resources projects are seen as among the most discretionary of federal expenditures. Unlike the entitlements programs (e.g., Social Security) which require positive governmental action to achieve cuts, all that is necessary to achieve reductions in expenditures is inaction. By the late 1970s and early 1980s, appropriations for these projects were increasingly being handled by Congress with continuing resolutions. Under continuing resolutions federal agencies are allowed to continue spending for ongoing programs at the previous year's rate, but no new construction initiatives are possible.

With arrival in the early 1980s of projected budget deficits of $200 billion per year and more, the prospects for finding majority support in Congress for major new water resources projects became even more doubtful.

In this context, the second factor contributing to the funding stalemate--the high cost of many proposed construction dredging projects, takes on increased importance. Table 13 (Appendix G) suggests the magnitudes of these costs. Note that five of the seven largest ports (in tons handled) have proposals for new construction dredging. Estimates by the Corps of Engineers for each of these construction projects range from $371 million to $479 million. In a context of large deficits, new project initiatives of this size have raised serious questions. Although the Corps' benefit-cost analyses for each of the projects show a positive benefit-cost ratio, these new projects require additional appropriations, and therefore, they represent absolute increments in the federal deficit.

A third factor frequently identified as contributing to the funding stalemate is self-amplifying. The longer the funding stalemate continues, the more difficult it becomes to reinitiate the old process. A brief recapitulation of the traditional time sequence associated with port dredging is useful in clarifying this third factor.

New construction dredging projects completed during the 1970s were regularly initiated 20 years earlier. Although there was widespread criticism of the long lead times required, they had significant benefits for the traditional, congressional funding process. First, the funding costs were spread over many years, so the costs for any individual project for any given year were relatively low. Second, project costs started at very low levels and incrementally increased, so that the high costs were incurred at the end of a long period. Little opposition could be mobilized against the low dollar expenditures associated with initial feasibility studies which themselves might be spread over several years, and so it was with initial design study costs. By the time initial construction costs were called for, a different and seemingly powerful logic had intervened. It was two-fold. First, "This project has been studied and reviewed over a very long period of time and repeatedly found to be justified." Second, "It would be a substantial waste to have invested all of these initial funds and not complete the project."

Given the expectation that authorization of a major new construction dredging project involved long years, individual ports were accustomed to an incremental process. So long as the expectation was that new construction dredging would ultimately be authorized but would inevitably take long periods of time, the mechanism for quid-pro-quo negotiations existed in the congressional appropriations process. Coalitions for support of individual projects could be built up in Congress based on the assumption that "if you support my beginning this year, I will support your beginning next year...or at some future date." The long incremental authorization and funding process, on the one hand, never provided any port with all it wanted, but on the other hand, provided most ports with some of what they wanted on a consistent basis. Most years, ports could expect that some incremental step would be taken toward satisfying their aspirations.

After more than a decade in which no significant new construction dredging has been initiated, the expectations of the various ports are very different. Rather than new construction proposals entering into an ongoing stream or process of authorization and appropriations, a number of major port construction projects are lined up together at the starting gate. With future funding uncertain, there is little incentive for any port to agree to be anything other than first out of the gate. In sum, the port community no longer appears capable of presenting a common front on priorities to Congress.

Finally, many believe that the erosion of the consensus on how ports should be funded is, in part, a reflection of changing social attitudes and values. This argument suggests that opposition has grown to "big government" and to governmental activities which are perceived as using general tax revenues to benefit narrow economic interests. One response to the concern with "big government" is that many governmental activities would more appropriately be carried out in the private sector. Such ideas are reflected in proposals that call for the private sector to take over some of the services provided by the U.S. Weather Service.

Where it does not appear feasible to transfer activities completely to the private sector, proposals are made to introduce market-like control mechanisms into the public sector. These views are summarized in a letter from five prominent economists to the chairman of the Senate Environment and Public Works Committee (October 18, 1983): "new or increased user fees for ports and inland waterways, market-based pricing for hydroelectric power..., and increased cost sharing and financing by non-federal entities for all federal water resources" would "lead to a more rational federal water resources policy....Economists have shown that economic benefits are enhanced when a project's beneficiaries pay in accord with the costs they impose and the benefits they receive." The concept is generally reflected in proposals for raising port dredging money with user fees.

These considerations, and others, have contributed to the demise of the consensus in Congress that general funds should be used to fund port dredging. What appears clear is that before new dredging programs can be undertaken, the old consensus must either be reestablished or a new method of funding must be found. The rapid rise in coal exports from the United States in 1980 resulting from the combination of the Iranian oil disruption, political difficulties in Poland and labor problems in Australia, together with the backlog of proposed port projects gave major impetus to the search for a new funding consensus.

NECESSARY ELEMENTS FOR CONSENSUS

Before a new funding consensus can be established for port dredging, and therefore, before stability can be returned to the port funding process, five interrelated issues will need to be resolved. First, a formula must be established which determines who will pay for dredging. At its most general level, this choice involves deciding whether, and if so, what portion of funding will be paid for from general tax revenues. If that portion is anything less than 100 percent, then a determination will have to be made concerning who will be required to pay and what mechanisms will be used to collect the funds.

Second, it will be necessary to determine who will collect the revenues. Revenue-raising responsibility can either rest totally with the federal government, totally with individual ports, or it may be shared. Shared responsibility ("cost sharing") implies that some portion of the revenues will be raised by the federal government and some portion by the individual ports, with an obviously important issue being what the relative portions are.

Third, any new consensus that changes the arrangements concerning who pays for dredging and who collects the revenues will likely be connected with new arrangements for revenue allocation. The traditional process of congressional negotiations associated with annual appropriations bills will be questioned if such changes occur. Depending on how revenues are raised and who raises them, the payers and collectors of these revenues will likely insist on a process that assures a return on investment in a rapid manner that at least partially assures that major payers will be the major recipients of the benefits.

Fourth, any changes to the above three sets of arrangements have the potential for changing the management and implementation role of the U.S. Army Corps of Engineers. For example, if the choice were for individual ports to raise all of their own funds for dredging, or a significant portion, it is quite possible that the individual ports would seek to exercise management control.

Finally, the evolution of a new consensus on port funding will doubtless require modifications in, at a minimum, existing permit approval processes. These processes apply to the dredging projects initiated by ports (or other local interests) that the ports will fund. Most proposals for federally funded new construction dredging will involve port-funded dredging as well, and depending on the funding mechanism finally selected, the permit process could take on added importance or replace the Congressional process. The ports have in recent years pressed for changes in this process, arguing that it is indefinite and without fixed limits (see Chapter 7). If required to raise and spend more for port dredging, the ports will likely demand more certain time horizons for investment decisions. The goal will be to accelerate the approval process and for that to occur, both legislative and regulatory changes may be necessary.

FUNDING ALTERNATIVES

Three options are possible as funding sources for dredging. The first, the traditional source, is general fund revenues. The second makes use of existing revenue sources at individual ports. The third involves new federal authorization for levying user fees or specialized taxes. In this latter connection, the right to levy user fees or other specialized taxes could be given either to the federal government, or to the individual ports, or to both.

GENERAL FUND REVENUES

Although proposals aimed at finding a new consensus on funding range across the spectrum, only one, the Reagan Administration's 1982 proposal, would totally eliminate general fund revenues as a source. However, if precedent is in any sense suggestive, the likelihood is high that general fund revenues will continue to pay some portion of dredging costs. For example, in 1982, according to the Energy Information Agency, slightly over half of the $23.3 billion in federal expenditures for transportation came from general fund revenues (Energy Information Administration, 1983). Of that amount, roughly $337 million was expended on U.S. Army Corps of Engineers operations and maintenance dredging at ports. Based on the transportation precedent and the fact that, with the exception of the Reagan Administration's, all of the legislative proposals aimed at breaking the funding deadlock have included general fund revenues, it appears likely that any new consensus on funding will involve general tax revenues covering a portion of dredging costs.

PRESENT PORT REVENUES

Although some proposals aimed at breaking the funding deadlock have involved a combination of general fund revenues and revenues raised through new federally authorized user fees, one option is to leave the ports dependent on only those revenues that they could raise under authorities now available to them. Ports have various sources of revenue. They range from wharfage and warehousing charges through revenues from non-port activities in those instances where port authorities operate airports and commercial properties, to state and local tax revenues. The capacity of ports to fund new dredging varies greatly depending on the volume of cargo they handle, the range of funding they have available, and the costs of both construction and maintenance dredging.

At one extreme are ports such as Los Angeles and Galveston, which faced with the present stalemate, are prepared to pay construction dredging costs. Los Angeles recently completed deepening of the harbor (in connection with federal dredging of the authorized project) to depths ranging from 45 ft to 51 ft at a cost to the port of $37 million. Galveston is committed to deepening its port to 56 ft at a cost of $139 million.

The Port Authority of New York/New Jersey has indicated to Congress that it is willing to advance the $110 million cost of a channel-deepening project, if assured of an expedited permit-approval process. By comparison, "Baltimore continues to insist on 100 percent federal funding for its project because it is the only port authorized to be deepened with federal funds" (Energy Information Administration, 1983).

At the other extreme, many small ports indicate that they are without the capacity to pay the costs of new dredging. A review of public statements by ports indicates that with very limited exceptions, some ports are either unwilling or incapable of paying for either routine maintenance dredging or new construction dredging until some new consensus has been formalized by legislative action.

Considerable variation exists concerning the ability of ports to pay all or some share of routine maintenance dredging and new construction dredging. How ports are organized and how they perceive their roles influences the rates they charge for port services. Some, for example, are partially supported by state or local revenues, owing to their importance in local and regional economies and their competitive status with other nearby ports. Other ports contribute revenues to state or local governments. The range of port charters and institutional identities represent various interpretations of their mixed public- and private-sector nature. Ports vary in the degree of control exercised by state or local governments, and ports have differing recourse to state and local bond issues (as indicated in a succeeding section).

Reviewing the existing rates levied by various ports in the United States (or charges they are now allowed to levy by federal law) evidences these disparities. Wharfage charges per ton of general cargo average $3.50 to $4.00 on the West Coast; $1.00 to $1.30 in the Gulf of Mexico, and $1.45 to $1.65 on the East Coast. Land leases and other port charges exhibit a comparable range. Because the range of costs among geographical regions in land, labor, and construction are not as great as the range of port charges, the principal reason for the remaining disparities would seem to be institutional differences.

USER FEES

New federally authorized user fees have been the most frequently proposed vehicle for building a consensus on funding. Substantial precedent exists for establishing port user fees as revenue sources. As previously noted, slightly less than half of the $23.3 billion expended by the federal government in 1982 for transportation was provided by user fees. A user fee in the form of an excise tax on motor fuels provides the revenues for the Federal Highway Trust Fund. Taxes on passenger tickets and other items provide the revenues for the Airport and Airway Trust Fund, and beginning in 1980, the Inland Waterways Trust Fund began receiving revenues from a fuel tax levied on barge operators. (The Inland Waterways Revenue Act of 1978 established a fuel tax of 4 cents per gallon of fuel for 1980, and 2-cent increases every 2 years ending in 1986 at 10 cents per gallon of fuel. Revenues in the trust fund are for new construction and major rehabilitation—e.g., of locks—on the inland waterways, but not for routine operation or maintenance.)

In its 1982 budget submission, the Reagan Administration sought to eliminate the use of general fund revenues for dredging and replace them with funds raised through new user fees. Under the Administration's proposal, port user fees (together with inland waterway user charges) were expected to raise $2.1 billion in revenues over the 1983 to 1986 period (Office of Management and Budget, 1981).

As interpreted in an Energy Information Administration (1983) report:

> The Administration's proposal was aimed at removing the federal subsidy from navigation programs, reducing the growth in federal spending, and moving toward a balanced budget. Besides reducing federal budget deficits, the justification for user fees rests on the efficient and equitable allocation of limited federal funds. In this argument, a user fee system becomes an efficient market test whereby only economically viable projects are selected out of a multitude of proposals. Port development yields significant benefits to port users who are not only able to pay but should pay for the benefits. User fees ease the burden on federal funds thus promoting more efficient and equitable allocation of these limited funds among competing purposes.

As noted in this statement, the Administration and other advocates of user fees attribute three distinct advantages to them: (1) new revenues, (2) economic efficiency, and (3) economic equity.

Faced with growing budget deficits and general resistance to increased taxes, governments at all levels in the United States have moved toward broader use of user fees. User fees offer the advantages of increased revenues while at the same time mobilizing minimum opposition. By tying user fees to the delivery of specific goods and services, payer opposition is diluted since the payers are also the beneficiaries. User fees are particularly efficacious revenue-raising instruments when the payer-beneficiary perceives a strong need for additional goods and services and is faced with the choice of either paying a fee or not receiving the needed good or service. Such appears to be the situation for those interests pushing vigorously for additional dredging. In theory, user fees are most attractive when they are levied such that the interests with the most pressing need pay the highest fees. In practice, however, that formula usually must be modified to take into account the capacity of the payer to pay. Part of the burden may be allocated to a broader set of users at high-value or high-volume ports (or both) without vigorous opposition if the absolute cost per ton or per dollar represents a small or insignificant addition to existing costs. However, even though the absolute cost per ton or per dollar may represent only a small addition to a large port's charges, it is very difficult to convince the larger ports (and their users) that they should be paying an amount in excess of that required for their own new construction or maintenance dredging if the surplus is funding the dredging of another port or ports. While specific commodities have been the focus of some proposals for new construction dredging, it must be remembered that all coastal ports compete with one another for general cargoes (OCP, or "overland common-point" cargoes).

Second, some advocates of user fees explain their support by arguing that such charges provide the public sector a vehicle for simulating market-like allocation decisions. That is, user fees, appropriately formulated, are said to bring standards of economic efficiency into public sector choices. Where users are required to pay the full costs for public goods and services, fees provide users with accurate information with which to evaluate their options. For example, if shippers are required to pay both the full cost of port deepening and the full cost of lightering/topping-off, they will choose the most economically efficient of the two options. If that were topping-off, the pressure for some dredging needs would presumably no longer exist. In a similar vein, user fees are said to provide public managers with information which will allow them to evaluate both the quantity and the quality of the goods and services they provide. In theory, then, appropriately formulated user fees would ensure adequate port capacity while at the same time protect against excessive capacity. Third, user fees are advocated as vehicles for achieving equity. In theory, user fees require those interests who benefit from public goods and services to pay in proportion to the degree they receive benefits. Alternatively, those who don't benefit don't pay.

In practice, establishing systems which achieve the stated benefits of user fees has turned out to be extremely difficult. In the case of port dredging, some interests simply reject the notion that standards such as efficiency and equity should be applied. Quite clearly, efficiency and equity standards applied in any pure form would have the result of closing certain ports. Where user fees threaten the existence of a port, efficiency and equity arguments have little appeal.

An additional complication is the wide range of differences among the ports of the United States in physical characteristics. Those having naturally deep water and those having lower maintenance dredging requirements than other ports are opposed to user fees, particularly if these fees are assessed on a nationally uniform basis. Thus, the principle of those receiving the benefits paying the bill cannot be actualized through a uniform nationwide fee. For example, the Congressional Budget Office (1983) estimates that a system of full cost recovery from user fees for small ports (less than 100,000 tons per year) would require those ports to charge a user fee of $90 per ton to recover all the costs associated with operations and maintenance dredging now provided by the Corps of Engineers, but for large ports (over 10 million tons per year) the charge would be $.20 per ton.

Even where there may be agreement in principle to the application of efficiency and equity standards, there is seldom agreement on the data base that should be used for calculating dredging needs and the establishment of a user fee system. First, in the case of new construction dredging, a central justification is always based on some projection of future need or opportunity. That, in turn, rests on projections of the future growth of the world economy and future trade patterns. We've already noted that there is little agreement to be found on these projections. What is clear is that if user fees were required to pay the full cost of construction dredging, they would likely be high at the beginning of the period of amortization and decline over time as the volume or value of cargo over which they were distributed grew. If the user fees are high on a per unit of cargo basis at the beginning, however, that may have the effect of diverting cargo to other, lower-cost ports, and the projected volume would never be achieved.

There are other practical matters that weigh against the application of users fees to specific beneficiaries. Ports that have a number of terminals essentially in competition with one another as well as with terminals in other ports are already faced with equity problems in port pricing. As a general example (exclusive of cargoes handled), if a port has two similar terminals, one constructed 15 years ago and one constructed in today's market, the first may have been built at a total construction cost of $10 million and the second at $50 million. The difference in dollar amounts may provide a 5 percent to 10 percent increase in efficiency; that is, the difference in cost is greater than the difference in efficiency or cargo-handling capability. The terminal operator in the 15-year-old terminal is operating from the facility that has been amortized, but the operator in the new facility has $40 to $50 million to amortize. In practice, this essential difference in the competitive status of the two operators is resolved by the port: the port's demands for funds to support operations, maintenance, and development are calculated and distributed as equitably as possible over all competing terminals. As a result, the operator of the 15-year-old terminal pays somewhat more and the operator of the new terminal somewhat less than their respective amortization costs. An analogous situation would be created by new construction or maintenance dredging if the port pays the costs. It is not unlikely that if user fees for a deeper channel or other facilities and associated terminals are assessed to only the users of these facilities, the economic justification for the projects will appear questionable.

Another issue of controversy revolves around the question of whether the users are the primary beneficiaries of dredging. The user fee concept rests on the assumption that the primary beneficiaries are the identifiable users of a publicly provided service. In the case of dredging, proposals for user fees generally call for a levy against ships. Shipping interests regularly argue that these proposals require them to carry the revenue burden for dredging, while in fact there are many other beneficiaries. Some economic analyses argue that in the case, for example, of coal exports, railroads, miners, and retail and wholesale businesses in mining regions would also be major beneficiaries. Others carry the argument much further, suggesting that deeper ports would have significant economic multiplier effects for the national economy. A recent study for the Corps of Engineers Los Angeles District (Data Resources, Inc., 1983) projects the economic benefits--in terms of total industrial production---from channel dredging and landfill developments in the ports of Los Angeles and Long Beach from the present to the year 2020. The following table gives a brief summary of the study's projected benefits in direct and indirect revenues and geographical component. While the direct effects are concentrated in the immediate area of the port, the indirect revenues are distributed across the country. Bushnell, Pearsall, and Trozzo, Inc. (1983) find similar indirect effects resulting from a uniform fee for maintenance dredging.

Regional Distribution of Industrial Production Revenues from Channel Deepening and Landfill Developments, Los Angeles/Long Beach, California: 1983-2020 (total cumulative 1983 dollars in millions)

	Los Angeles County	Five-County Region	California	Six-State Region	National
Direct	1,056.8	1,396.9	1,682.0	1,697.8	1,832.8
Indirect	179.3	249.7	456.2	711.0	4,830.7
Total	1,236.0	1,646.6	2,138.2	2,408.8	6,663.5

The argument at the national level is that (for example) increased coal exports would reduce our balance of trade deficit, lower unemployment (and with that, the need for federally funded social services), and so on, to the end that new construction dredging of ports should be paid for from general fund revenues since the nation benefits. In sum, in a complex economy it is simply impossible, so this argument goes, to sort out the primary beneficiaries of port dredging, with the result that levying fees against any specific set of users is inherently inequitable.

Despite these difficulties, some analysts believe that the answer is institutional. That is, the way to create a market-like situation is to create competitive ports by withdrawing all federal support for dredging whether for new construction or maintenance. Under this arrangement, Congress would authorize the ports to levy user fees in any way or at any level they were to determine. Critics of this approach note that ports are inherently creations of government and cannot be made to operate as purely private-sector organizations. They note that the legal and institutional character of U.S. ports varies-- from ports managed by state governments to those that are the creations and responsibilities of city governments, and others run by relatively autonomous port authorities. These differing situations have potentially very different consequences for port financing.

While there are many sound reasons for enhancing the market-like conditions of port operations to make the ports more profit-oriented, an underlying issue remains their public character; phrased differently, to whom do the ports really belong? The importance of international trade to the domestic economy of the United States, the dominance of oceanborne shipping in international trade, and the role of the ports in national security and defense suggest that the ports are national assets.

Where ports are managed by agencies of the state government, or alternatively, city governments, they may be able to borrow money at reduced rates using the full faith and credit of either the state or the city. Where the ports have substantial political leverage, it is reasonable to assume that general tax revenues from either the state or the city might well be used to subsidize dredging costs. In the cases of those ports run by authorities that also manage airports and other commercial facilities, the possibility exists that profits from these non-port activities will be used to subsidize dredging and therefore potentially give those ports the ability to charge lower user fees with the associated competitive advantage.

Financing arrangements for port dredging can make massive differences in the user fees that must be charged. For example, the ability to amortize capital costs over 50 years versus half that time or less can significantly affect financing costs. Similarly, the ability to borrow low-interest or no-interest money versus market-rate money can make a decisive difference. These factors are heavily influenced by the legal and institutional structure of the port and vary from one port to another.

Finally, the form of the fee or tax obviously has very different implications for different interests as well as for the costs of collecting the fee. The key point is that while analysts may assess user fee options based on abstract standards, those interested in port dredging are concerned about who benefits and who pays. In the context of support for various port funding proposals, the divisions are clear. Low-volume ports with high-cost dredging requirements support a uniform national user fee. Such a fee is attractive because it requires high-volume, low-cost ports to provide them subsidies. For these very reasons, high-volume, low-dredging-cost ports prefer a user fee which is port specific.

In the same vein, shippers of high-value, low-volume commodities favor a tonnage-based fee, whether national or port-specific. Alternatively, high-volume, low-value shippers prefer an ad valorem tax, whether levied on a nationally uniform basis or on an individual port basis.

Finally, there are a number of issues that revolve around the efficient management of any user fee collection system. From the point of view of economic theory, the most efficient user fees are those that reflect the marginal uses. If user fees are to closely simulate a free market, the essential purpose is to assure that the marginal benefits of any new investment will be greater than the marginal costs of that investment. In the changing environment of international economies and trade, that requires detailed collection and analysis of data and a great amount of flexibility would have to be granted to those setting the fees. Collection, analysis, and management of this kind of data is, in and of itself, high cost, and a management system with this kind of flexibility also requires that governing bodies give a great deal of discretion to administrators to allow them to act quickly. The general pattern in the United States, however, is to resist building the kinds of large administrative systems necessary to manage user fees that are responsive to marginal costs and benefits, and similarly, legislative bodies generally resist giving broad discretionary authority to administrative organizations.

Alternatively, the lowest cost and simplest user fees to administer are those that are broadly based and therefore incapable of distinguishing between marginal and non-marginal costs and benefits. As an example, the federal tax on motor fuels is broad and easy to collect. The nine cent per gallon tax on motor fuel is referred to as a "highway tax." It is actually collected from a small number of refiners and distributors and is cheap and easy to administer. Alternatively, it does not reflect any difference between cars and trucks and the amount of damage they do to highways. It therefore does not meet the more refined definitions of efficiency and equity which are frequently used to justify user fees.

Two conclusions must flow from any review of the user fee debate. First, user fees are being proposed as a vehicle for finding a new national consensus on port funding. Second, there is no agreement on the role user fees should play or how they should be applied. If user fees turn out to be the structure around which a new funding consensus is evolved, it will be because they serve as a mechanism for evolving compromises. The conflict over funding is a conflict of values and goals, a conflict about who pays and who benefits. Those conflicts can only be resolved in the political process with political compromises.

Intertwined with the proposals for alternative sources of revenues are various proposals concerning who should collect the revenues. Three categories of options exist. The first would have all revenue raised by the federal government. Clearly, if all revenues were to continue to be derived from general taxes, the dominant federal role would remain the same. Similarly, the Reagan Administration's proposal would retain the same federal dominance but would derive all dredging revenues from a new source, a user fee.

Second, at the other extreme, individual ports could be made responsible for raising all revenues. This arrangement would exist if, on the one hand, ports were required to raise revenues from existing sources, or, on the other hand, if the federal government passed legislation authorizing ports to establish new taxes or user fees to be determined by individual ports.

The third option would have both the federal government and individual ports raising some portion of the revenue. A review of the proposed legislation dealing with dredging indicates that every proposal with the exception of the Administration's first (100 percent cost recovery, later revised to 70 percent) calls for some form of joint funding by the federal government and the individual ports. Stated differently, most efforts to build a new consensus on port funding have included what are known as cost-sharing arrangements.

Cost sharing between the federal government and state or local governments has a long history. Historically, cost-sharing has been an instrument used by the federal government to induce state and local governments to carry out new activities. The pattern has been for the federal government to establish programs which commit federal funds to paying for some percentage of given activities if state and local governments match those funds. Cost-sharing programs, for example, have been instrumental in the federal highway program; in the case of interstate highways, the federal share is 90 percent and the state share is 10 percent. Similar cost-sharing arrangements have been the instrument for initiating and carrying out a wide range of programs from environmental enforcement activities to a broad set of social welfare programs.

In the case of port dredging, the motives behind cost sharing are different. Cost-sharing proposals in this sector have as their goal getting the ports to assume responsibility for a greater portion of funding for an activity which traditionally has been fully funded by the federal government (except for local sponsor costs). That is, cost sharing is a way to transfer what have traditionally been federal responsibilities and costs to the state and local level.

To the extent that cost sharing is attractive to individual ports, it is because such cost sharing is seen as a way either to increase the absolute level of federal funding by offering a formula which would reduce the percentage or the proportion of federal funding, or to achieve some other benefit such as fast tracking of required regulatory review (discussed in the section, "Non-Funding Issues"). Given the funding stalemate, many observers believe that the only way to increase federal dollars is for the individual ports to assume some greater portion of the costs. Cost sharing, then, may be attractive to the individual ports as a vehicle for prying loose additional federal dollars to pay for new construction dredging, or increased maintenance dredging, or for both.

It must be emphasized that the cost-sharing concept does not inherently provide any answers to the question of who will pay. The federal government's portion of any cost-sharing formula could come from either general fund revenues or from a new tax or user fee. Similarly, in the case of individual ports, revenues for cost-sharing could come either from existing sources such as wharfage, dockage, stevedoring, and harbor transfers or from state or local tax revenues, or they could come from new federally authorized user fees.

The key point is that although cost sharing is an integral part of most proposals aimed at finding a new consensus on funding, the principle does not by itself imply anything about the source of revenues.

This point can be illustrated by looking at three simple hypothetical alternatives. If a 50-50 cost-sharing arrangement for all dredging were put in place which required the vessels using each port to pay the full costs of that port's operations and maintenance dredging, the average charge for small ports (under 100,000 tons) would still be $90 per ton of cargo, while for large ports it would be $.20 per ton (Congressional Budget Office, 1983). The only qualification is that if both the federal goverment and the local ports were charging a tonnage fee as the sole basis for raising their share, the cost of two administrative structures to collect those fees might very well make the charge even higher.

Alternatively, if there were a 50-50 cost-sharing arrangement, with the federal government levying a uniform national user fee, the federal levy needed to cover just operations and maintenance would be an average 12.7 cents per ton at all ports in the United States, while the local user fees would range between an average $45 per ton for small ports to an average $.10 per ton for large ports. (Congressional Budget Office, 1983).

Finally, if the cost-sharing arrangement took all of the federal share out of general fund revenues and the local ports had to pay their portion of a 50-50 split from user fees, the arrangement would be the same as above for the individual port with no user fees charged by the federal government. In sum, large ports would still charge an average $.10 per ton and small ports an average $45 per ton.

The range of proposals for cost sharing is potentially infinite. What is clear is that if cost sharing is to be one of the elements of a new funding consensus, it will require the establishment of a formula that is broadly acceptable, and achieving acceptability will involve a complex set of political compromises.

Most of the legislative proposals calling for some kind of cost sharing have sought to build consensus on port funding by establishing some kind of ceiling on how much money individual ports would have to pay in an effort to protect ports against excessive costs. Two different ceilings have characterized these proposals. One approach involves "grandfathering" depths, and the other establishing a tonnage-fee ceiling. For example, House and Senate bills considered by Congress in 1983-1984 (H. R. 3977; S.1389) set 14 feet as a threshold. That is, small ports with depths of 14 feet or less would not be required to pay any dredging costs. Another House bill used a 45-foot depth which retains the traditional federal role of paying the full cost of navigational projects, including both new construction and maintenance dredging to 45 feet of depth. Only improvement projects for dredging in excess of 45 feet would be subject to cost sharing. An alternative Senate bill proposed the adoption of a port-specific tonnage fee for maintenance dredging but set a maximum tonnage fee so that no port would have to pay an excessive amount. This bill, however, retained the port-specific cost recovery scheme for any new construction dredging.

The most recent legislative acts and also the most nearly successful were two bills introduced in the 1984 legislative session. HR 3678, a water resources bill, passed the House by an overwhelming 7-1 margin. It contained many new authorizations and a large number of deauthorizations of projects that had been on the Corps of Engineers program for several years without action. This bill contained what has now become a widely accepted principle for operation, maintenance, and new construction; that is, federal responsibility to 45 ft in depth and cost sharing for additional increments of depth. The bill also provided for a revolving fund to finance federal participation in both new construction and maintenance dredging. The fund would have a $2 billion reserve, all allocated from customs collections, which now amount to more than $7 billion annually ($6 billion/year from the coastal ports). Member ports of the American Association of Port Authorities overwhelmingly supported this bill.

Senate bill 1739, which contained some of the features of HR 3678, was more restrictive on the handling of user fee collection and disposition. It did not contain as large a number of authorizations as the House water resources bill but did provide some port-specific flexibility to ports in the authorization and assessment of user fees; nevertheless, language concerning the collection of user fees from beneficiaries made the bill unworkable in the opinion of most of the major ports in the country. Senate Bill 1739 did not reach the floor of the Senate. Both bills were attached to their respective House and Senate continuing resolutions prior to the passage of the 1984/85 federal budget. Conflicts between House and Senate members over provisions of the two bills and strong opposition from the Office of Management and Budget resulted in both bills' being dropped from the continuing resolutions; thus, there was no action on water resources legislation in the 98th Session of Congress.

As this brief review indicates, of the bills proposed and seriously considered by Congress to overcome the funding barrier, none was primarily concerned with equity and efficiency.

ALLOCATION OF REVENUES

Any changes in either the sources of revenue (e.g., new taxes or user fees) or in who collects those revenues (e.g., cost sharing) will likely create major pressures for modifications in the traditional processes for allocating revenues. The traditional process, in which Congress allocates General Fund revenues on a project-by-project basis in authorizations and appropriations, has involved long lead times.

If new user fees were established, the collectors of those fees would doubtless seek to assert a greater role in the allocation process, both in an effort to accelerate dredging activities and to assure that the ports paying the primary portions of the fees would receive priority attention in the allocation process. Similarly, cost sharing, which makes individual ports collectors of the revenue, would likely see those ports demand some role in allocating the revenues.

A variety of allocation mechanisms can be hypothesized. One option might involve the establishment of a national port plan. Such a plan in its extreme form might identify a limited number of ports that would have first priority for deep-draft capability (more than 45 feet) but it would clearly be extremely difficult to establish in the political context of the United States.

Another approach might involve the establishment of a trust fund whose general allocation criteria would be laid out by Congress with an Executive Branch agency, such as the Corps of Engineers, formulating the detailed criteria and allocating the funds on a port-by-port basis. The model would be the highway or the airport and airways trust funds. Ports have generally expressed skepticism about the administration of such a fund, skepticism generated in part by the participation of many ports in the airport and airways trust fund.

A third approach might be some kind of competitive bid situation in which those ports willing to participate in cost sharing would bid for first priority in revenue allocation. Under such a formula, the port willing to put up the largest percentage of matching funds, above some fixed percentage floor, would be given first priority for federal funds.

The key point about revenue allocation is that any new consensus on funding that changes the source of the revenues and the organizations that collect and dispense the revenues will likely require the evolution of a new consensus on how those revenues should be allocated. Without such arrangements, the possibility exists of having new revenue sources but being unable to allocate the resources. The consequence would be that funding barriers would remain.

MANAGEMENT AND IMPLEMENTATION OF PORT DREDGING

So long as general fund revenues provided the monies for both maintenance and new construction dredging, the Corps of Engineers managed all major dredging projects. Such projects now require congressional authorization and are funded on an annual basis similar to other federally authorized programs.

In the abstract, there are substantial advantages in having a single national management organization responsible for dredging. That is particularly the case where port dredging is interrelated with other social objectives. For example, in the case of the Port of New Orleans, the Corps has responsibility both for maintaining the navigational channels and for flood control. These two activities would appear to be nearly inseparable. Further, where the Corps has sole responsibility, procedures involved in carrying out dredging are the same nationwide. And finally, the Corps--with its broad base of experience and its research and development program--provides a technically competent organization.

As sources of funds and revenue-raising responsibilities change, the possibility exists that the Corps' role might change in fundamental ways. The range of possibilities is broad. At one extreme, individual ports might insist that they be the managers of their port dredging activity, contracting with private consultants and private dredging companies for all of the work, with the Corps' role being reduced to that of an approver of permits. Alternatively, the Corps might become a contractor to the ports for design, or it is possible that some ports would allow the Corps to play its traditional role for federal projects, the only difference being the source of funding.

The key point is that any change in funding sources and collection arrangements has the potential for requiring changes in traditional management procedures. Again, these changes, depending on how they are worked out, could themselves become barriers to timely port dredging.

NON-FUNDING ISSUES

Almost without exception, proposals for new funding and collection arrangements aimed at finding a new consensus on port dredging have involved calls for what has become known as fast tracking. Most of the parties interested in port dredging find the present 20-year (or more) lead time which has characterized completed projects to be unacceptable. So long as general funds were the source of dredging money, these long lead times served to provide a stable environment within which priorities for port projects could be evolved. With the arrival of the funding stalemate, however, demands for fast tracking have received increasing attention.

What fast tracking would involve seems to vary. Discussions pursuant to finding a new consensus have ranged from escaping the requirement for congressional authorization for new construction dredging to substantially accelerated permit approval for locally funded dredging and filling projects by the involved federal and state agencies. Fast-tracking--that is, reducing the lead times either for congressional project authorizations or for agency permit approvals-- becomes increasingly important with progressively lower federal funding for port dredging. When the federal government paid a major portion of the costs for new construction dredging, the time value of money did not become a major issue. However, if individual ports assume all or a major portion of the costs for new construction dredging, the present long lead times and the present and future time factors of inflation and money may change the balance of benefit-cost ratios. Certainly for dredging/filling projects necessary to the development of a new terminal, securing outside financing is likely to be the critical factor in proceeding with the project, and the ability to secure tenant financing will be impossible if the prevailing long lead times continue. The same considerations apply to projects involving dredging or filling that the ports have traditionally funded but that need permit approvals. For any major port-funded project, the single issue of greatest importance in the present open-ended and indefinite permit-approval process is the inability to fix the time horizon of decision making, or to identify an end-point.

The point to be emphasized is that any changes in any of the above four categories of issues will doubtless also require some kind of modification in, at a minimum, the dredging approval process. Without that, the possibility of a funding consensus appears dim, and without some agreement on fast tracking, all of the other points could be resolved and the nation would find itself with a continued lack of dredging.

CONCLUSIONS

Any new consensus on port funding that allows the funding barrier to be overcome will require the resolution of five issues: (1) the source of revenues, (2) who will collect the revenues, (3) how the revenues will be allocated, (4) who will handle the management and implementation of port dredging, and (5) the integration into this process of some kind of modification in approval processes that allows more expeditious initiation and completion of port dredging to occur. It is essential that the political process address all of these issues if a new consensus on port funding is to be found that allows the nation to overcome the funding barrier.

REFERENCES

Bushnell, Pearsall, and Trozzo, Inc. (1983), "Economic Effects of Levying a User Charge on Foreign and Domestic Commerce to Finance Harbor Maintenance," Report to the Economic Development Administration, U.S. Department of Commerce.

Congressional Budget Office (1983), Reducing the Deficits: Spending and Revenue Options (Washington, D.C.: Government Printing Office).

Data Resources, Inc. (1983), Los Angeles/Long Beach Landfill Development and Channel Improvements: An Economic Analysis of the Army Corps of Engineers Master Plan to the Year of 2020, Executive Summary (Lexington, Mass.: DRI, Inc.).

Energy Information Administration (1983), Port Deepening and User Fees: Impact on U.S. Coal Exports, Report No. DOE/EIA-0400 (Washington, D.C.: Government Printing Office).

Office of Management and Budget (1981), Fiscal Year 1982 Budget Reviews (Washington, D.C.: Government Printing Office).

7
The Institutional
Decision Making System

INTRODUCTION

Decision making for port dredging became a major concern for two reasons. One is discussed in detail in the preceding chapter: the paralysis of funding for traditionally federal dredging projects. The second, closely related reason is the frustratingly long time that now elapses in the decision making process for approval of traditionally local projects, and for bringing proposed federal projects to congressional consideration.

This chapter highlights constraints in the decision making process that pose problems for dredging projects. It investigates a frequently proposed answer to these problems--"fast tracking." The chapter closes with an assessment of prospects for accelerating decision making, and for bringing stability and predictability to the decision making process.

The institutional decision making process for port dredging is complex, cumbersome, unpredictable, and fragmented. It is the product of legislation and regulation accumulated over the past 150 years. As constituted today, the system requires or provides opportunity for participation by Congress, the courts, a large number of federal agencies, as well as state and local governments, and many interest groups. The interest groups engaged in particular decisions may be numerous and diverse--commercial and other entities associated with ports, shipping and transportation firms, environmental organizations, citizens groups, and other members of the local population.

The system's complexity reflects its need to address and manage a complex set of needs and concerns. No dredging project represents an unmixed blessing to all concerned and there may be many concerned. Dredging decisions must assess a great deal of sometimes conflicting data, and balance a diverse set of interests that are frequently vigorously advocated.

As described in preceding chapters, decisions involve: which ports to dredge, who will pay for the dredging, what the appropriate design of the port will be, how it will be dredged, where the dredged material will be disposed, how best to manage the environmental effects, and how to respond appropriately to the concerns and responsibilities of governmental organizations and non-governmental

77

interest groups. Many of these elements vary from one port to
another, and fluctuate with time. Decision making about port
development must resolve real issues, gather and analyze real data,
and find accommodation among conflicting interests.

In the case of major federal projects, this decision making process
may take as long as 22 years. For most local projects, the time is
generally shorter but still far too long from the point of view of
those proposing the project. Not surprisingly, then, a wide range of
interests concerned with port dredging have expressed growing
dissatisfaction with the decision making process.

This dissatisfaction has led to ever more frequent calls for what
has come to be known as fast tracking. Although fast tracking has not
been clearly defined, its advocates do agree that the objectives are
speed, predictability, and stability.

The important role and responsibilities of the U.S. Army Corps of
Engineers in all port dredging projects--whether federally or locally
funded--makes the federal government the focus of concern of those who
advocate fast tracking. The federal role in ports results from three
basic developments. First, the Constitution of the United States
prohibits discrimination among the nation's ports by the federal
government. Second, since the passage in 1824 of the General Survey
Act, the Corps has had primary responsibility to oversee or carry out
dredging for the nation's ports. The Corps' initial responsibility
was to ensure navigability. Some of this responsibility is now taken
by the U.S. Coast Guard (placement of aids to navigation, for
example). Ensuring navigability by dredging is still a responsibility
of the Corps. Third, during the late 1960s and the decade of the
1970s, a broad set of environmental legislation gave the Corps and a
variety of other federal agencies responsibility for assessing the
environmental consequences of dredging and other activities and
ensuring that those activities met standards adopted to protect the
environment. The key institutional consequence of this body of
legislation was to require that the Corps take responsibilities far
beyond navigation and to assure that it coordinate and cooperate with
a variety of other federal agencies as well as state and local
governments. The Corps, then, is the key and lead federal agency for
dredging activities irrespective of origin or funding.

FEDERAL VERSUS LOCAL PROJECTS

Federal projects differ from local projects in a number of ways.
Historically, the federal government has assumed responsibility both
for the construction and maintenance of major access channels,
maneuvering areas, and anchorages in the ports of the United States.
This has meant that the federal government both funds and manages
federal dredging activities. Funding for federal projects has
traditionally been provided in omnibus authorization and
appropriations bills enacted by the Congress every two years or so.
The projects pass through several phases ranging from initial

investigations to physical construction. Movement from one key phase to the next requires specific authorization and funding by Congress, and intermediate steps—consultation with other federal and state agencies, the public, and preparation of reports for successive approvals by higher levels of the Corps—might depend on annual appropriations. The evolution from initiation to completion of federal dredging projects is outlined in Table 14 (Appendix G) and mapped against time in Table 15 (Appendix G). The average time from initiation to completion is 21.6 years. Over half the time is consumed in the congressional processes of authorizing and funding the project.

Two facts need to be emphasized concerning Congressional decisions for port dredging. First, the choice of which projects to fund, the level of funding provided, and the speed with which decisions are made is a product of the traditional processes of congressional negotiation. Within our system of government, there is no way to establish external discipline on this process. Any acceleration of the rate at which Congress makes these decisions or any increase in the predictability of these decisions will be made by Congress itself. Second, congressional authorization and appropriations describe the physical dimensions of dredging projects. That is, Congress typically specifies channel widths and depths. This latter point is important because the Corps, in carrying out congressional mandates, must frequently operate within precise guidelines. This can become a serious problem in a process that takes more than 20 years: the needs of the port may change significantly in the meantime. In sum, Congressional port decisions can become operating strait-jackets.

The second major decision maker with regard to federal dredging projects is the Corps of Engineers. It is critical to understand that the Corps' responsibilities are divided between management and regulation. On the management side, the Corps plans and designs federal projects and may either contract with the private sector for dredging or may use its own dredges to carry out the work.

On a separate track, the Corps is the major regulatory agency for dredging. Operating under the guidance of legislation and regulations governing navigation, safety, and a broad set of environmental and public conerns, the Corps has responsibility for approving the acceptability of its own activities. For example, the Corps, in cooperation and consultation with a variety of other federal executive agencies and state and local agencies must approve or deny such specific activities as the dredging itself, the transport of dredged materials, and the location and ways in which dredged materials will be disposed of.

Local dredging projects are not funded by federal monies. Typically they are concerned with dredging (1) to provide access to shoreside facilities, (2) of berths, (3) landfill projects (or some combination). Local projects do not require congressional action, and in most instances, they are not managed by the Corps. Local projects, however, fall under the regulatory authority of the Corps. They must meet the navigational, environmental, and social requirements derived from the body of dredging-related legislation.

Basically, the same laws and regulations apply to both federal and local projects. The only distinction is that the Corps does not formally issue permits for federal projects while it must issue permits for local projects.

FAST TRACKING

The growing demand for shortening the time required to approve dredging projects and bring predictability and stability to the process applies to both federal and local projects. In the case of federal projects, there are three categories of decisions where the goal of fast tracking might be achieved. The first category involves congressional decisions. During the post-World War II period, the amount of time consumed in the congressional decision making process increased to the point that the various authorization and appropriation decisions consumed 12 to 13 years of a 22-year initiation-to-completion period. The opportunities for time saving in the congressional decision making process are substantial. A 1978 report by the General Accounting Office made a number of recommendations to reduce the time taken up in decision making. In making these recommendations, however, the General Accounting Office noted that their adoption would have the effect of reducing congressional control and oversight (General Accounting Office, 1978). There is no evidence that Congress is prepared to give up oversight and control authority. This point was underlined by the General Accounting Office in a follow-up study six years later, which found that the "process for Corps water resource projects has remained essentially unchanged since our 1978 report. The options...have not been adopted by the Congress. Thus the Congress maintains the same level of control and oversight over water projects" (General Accounting Office, 1984).

As discussed in Chapter 6, basic changes in the way federal projects are funded may result in changes in the character of congressional decision making. What seems clear is that fast tracking in Congress and the source of funding are inextricably intertwined. It seems likely that decisions on sources of funding will necessarily precede any changes in the way authorization and appropriations choices are made and the outcome at this point is unpredictable.

Alternatively, it has been possible to speed up Corps decision making with regard to its management of federal projects. In response to growing concern over ever-longer lead times, the Corps revised its procedures. The Corps reports an average savings of 1.7 years in survey work and 1.4 years in review time for projects having reached final approval stages in the 1975-1977 time period (General Accounting Office, 1984). There may be further opportunities to enhance the efficiency and speed of the federal decision making process, but given the prerogatives of Congress, fast tracking clearly has limits.

A review of the concerns expressed by those calling for fast tracking indicates dissatisfaction with a third category of

decisions--those associated with regulatory responsibilities. These
regulatory decisions apply to both federal and local projects. Delays
associated with permitting decisions for local projects are the major
concern. The increasingly long lead times associated with gaining
permit approval correlate with the passage of environmental
legislation beginning in the late 1960s and continuing through the
1970s. This legislation broadened the regulatory responsibilities of
the Corps to include a diverse set of environmental and other
concerns. Further, this legislation was responsible for bringing into
the regulatory decision making process a significantly expanded number
of participants.

THE REGULATORY DECISION MAKING SYSTEM

The number of laws, executive orders, and policies that are (or may
be) pertinent to the regulation of dredging projects is substantial
(see Appendix E). In addition, every responsible federal agency has
put in place a set of federal regulations as necessary to carry out
congressional mandates and executive orders. It is both beyond the
scope and the capacity of this study to investigate this whole complex
in detail. It is, however, necessary to characterize the regulatory
decision making system to gain an understanding of why decision making
takes longer now than in the past.

The committee has chosen to do this by looking at some of the
organizational and procedural consequences for particular agencies of
historical (but still active) and recent legislation: the Rivers and
Harbors Act of 1899, the National Environmental Policy Act of 1969,
the Fish and Wildlife Coordination Act of 1958, the Endangered Species
Act of 1973, the Clean Water Act, the Marine Protection, Research and
Sanctuaries Act of 1972, and the Coastal Zone Management Act of 1972.

The succeeding section illustrates the complexity, overlapping
jurisdiction, and requirements for coordination resulting from these
and related pieces of legislation.

AGENCY ROLES

The Corps in its role as lead agency with regard to both federally
funded and local dredging activities must consult at numerous points
with a diverse set of other governmental agencies in carrying out its
regulatory responsibilities. Something of the texture of that
consultation and coordination is suggested by looking at four areas of
responsibility: (1) environmental assessment, (2) approval of local
dredging activities, (3) approval of fill or disposal in U.S. waters,
or (4) approval of transportation of dredged materials for ocean
disposal. In the case of the last three items, local projects require
formal permits while for federal projects the same regulatory
decisions are required, but no formal permits are issued. The
succeeding description is primarily concerned with agency roles and
regulatory decisions in these four areas of responsibility for local

projects, but there are some differences in the processes followed for federal and local projects, and in the instances these are important, the differences are noted.

U.S. Army Corps of Engineers

The Corps (generally acting through its district engineers) is required to follow procedures and prepare documentation established by the President's Council on Environmental Quality (CEQ) for carrying out the mandates of the National Environmental Policy Act (NEPA). In the case of major federal projects with potentially significant environmental implications, the Corps districts have a mandatory responsibility to prepare a full-scale Environmental Impact Statement (EIS). The initiation of local projects is signaled to the Corps by a permit application: the district engineer may initiate either an Environmental Assessment (EA) (a review to determine whether a full-scale Environmental Impact Statement is needed) or an Environmental Impact Statement (EIS), unless the proposed activity falls within a predetermined categorical exclusion. The objective of the environmental assessment process required by NEPA is to ensure that decision makers have available a broad overview of the systemic environmental effects of the proposed dredging activity.

The three specific approvals identified are handled through individual permit application approval procedures. Under authority derived from Section 10 of the Rivers and Harbors Act of 1899 (33 U.S.C. 403), the Corps has responsibility for issuing or denying permits for construction or other work in or affecting the nation's navigable waters. Under Section 404 of the Clean Water Act (33 U.S.C. 1251), the Corps has responsibility for issuing or denying permits for the discharge of dredged or fill materials in U.S. waters. Under Section 103 of the Marine Protection Research and Sanctuaries Act of 1972 (33 U.S.C. 1401), the Corps has responsibility for issuing or denying permits for the transportation of dredged material for open-ocean disposal. Permit-issuance authority is delegated by the Corps to its district engineers for local projects (with provision for higher-level review if the applicant seeks it). Before permit decisions can be made, however, the authorized officials of the Corps must consult with a wide range of federal, state, and local resource agencies and must provide for a public interest review. As these actions are taken, and comments are sought and received, the district engineers are authorized to add, modify, or delete special conditions (for example, actions to mitigate adverse environmental effects) under the Corps' broad responsibilities to protect the public interest.

The point to be noted here is that Corps regulations reflect the fact that each port situation is unique. District engineers are therefore provided with the opportunity to respond to those unique conditions. On the other hand, these special-condition options give the Corps a great deal of flexibility, and give to other agencies the means for insisting that permit approval be subject to special conditions they believe to be desirable. In practice, the Corps acts on special conditions only after the port and various interested organizations and parties have reached consensus. In this connection, the Corps normally consults with the U.S. Fish and Wildlife Service, the National Marine Fisheries Service, the Environmental Protection Agency, and a range of state and local organizations.

Finally, it is important to note that the port (or the Corps, for a federal project) must secure state certification that the project complies with applicable state water quality standards.

U.S. Fish and Wildlife Service (Department of Interior)

The mission of the U.S. Fish and Wildlife Service (USFWS) is to conserve, protect, and enhance fish and wildlife and their habitats for the continuing benefit of the people. The principal responsibility and authority is for migratory birds, threatened and endangered species, certain marine mammals, international resources, and wildlife on land under USFWS control. The Fish and Wildlife Service has responsibility for reviewing and consulting with the Corps on permit applications and environmental documentation pursuant to the provisions of the Fish and Wildlife Coordination Act of 1958 (16 U.S.C. 661); Migratory Bird Conservation Act of 1928 (16 U.S.C. 715), and international treaties; the Endangered Species Act of 1973 (16 U.S.C. 153); and National Environmental Policy Act of 1969 (42 U.S.C. 4321). Damage to wildlife resources from proposed projects and any possible routes to mitigation of that damage must be considered in the Corps' public interest review and in its environmental assessment procedure. Under authority gained from the Endangered Species Act, the Corps may not approve permits until it has received a "no-jeopardy" biological opinion from the Fish and Wildlife Service.

National Marine Fisheries Service (National Oceanic and Atmospheric Administration, Department of Commerce)

The role of the National Marine Fisheries Service (NMFS) in reviewing Corps permit applications or proposed federal dredging projects results from its responsibility under the Fish and Wildlife Coordination Act* for determining the probable effect of the projects on marine, estuarine, and anadromous or commercial fishery resources and their habitats. Specific consideration must be given to fish and shellfish resources, the presence of endangered fishery resources, and the biological significance of affected areas.

---

*Certain responsibilities of the U.S. Fish and Wildlife Service (in particular, those of the preexisting Bureau of Commercial Fisheries) were transferred to the National Marine Fisheries Service in 1970.

U.S. Environmental Protection Agency (EPA)

EPA's role in the Corps' permit process and in federal dredging projects stems in part from the Clean Water Act. This law designates EPA as the administrator of the act and states the objective of restoring and maintaining the chemical, physical, and biological integrity of the nation's waters. EPA has responsibility for developing and publishing guidelines for the discharge of dredged materials in the waters of the United States. Under this authority, EPA reviews Corps permit applications and Corps projects to ensure they adhere to the guidelines. EPA may veto Corps permits or proposals by prohibiting or restricting the use of any disposal area in inland waters of the United States if it determines that the discharge of such materials will have an unacceptable, adverse effect on municipal water supplies, shellfish beds, fishery areas, and wildlife or recreational areas (40 C.F.R. Part 230).

Under Section 103 of the Marine Protection, Research and Sanctuaries Act of 1972, EPA has responsibility (1) for developing environmental criteria; (2) designating approved open-ocean disposal sites; and (3) ultimate veto powers in permit approval. The Corps must consult with EPA to determine compliance with established guidelines. EPA has the ultimate authority to prevent issuance of an open-ocean disposal permit if the agency determines it will have unacceptable adverse environmental consequences.

Article IV of the Convention on the Prevention of Marine Pollution by Dumping of Wastes and Other Matter--better known as the London Dumping Convention (LDC)--imposes additional restrictions on the ocean dumping of "waste or other matter," including dredged materials. Most importantly, the LDC prohibits the dumping of materials containing certain constituents (specified in Annex I) when present as other than "trace contaminants" and when not rapidly rendered harmless after disposal. The LDC allows dumping permits to be issued only after consideration of all the factors listed in Annex III -- the characteristics and composition of the matter to be dumped, and off the dumping site; method of deposit; and general considerations and conditions. The Marine Protection, Research and Sanctuaries Act was amended by Congress in 1974 to require that EPA consider the LDC's standards and criteria in establishing or revising domestic ocean dumping criteria. In recent years, the International Maritime Organization (IMO), which administers the LDC, has had under consideration proposals from the International Association of Ports and Harbors to define certain "special care" measures, some of which show great promise, and which recent tests indicate might allow even highly contaminated dredged material to be safely disposed of at sea (see Chapter 9).

State Departments of Fish, Game/Wildlife

State departments of fish and game are trustee agencies having jurisdiction by state law over the fish and wildlife resources of

their states. Like the USFWS and the NMFS, the role of these state
agencies in review of permit applications and environmental documents,
as well as consultation about actions that may be required, is
authorized by the Fish and Wildlife Coordination Act, the Endangered
Species Act, and the National Environmental Policy Act.

State (and Regional) Water Quality Control Boards

State water quality control boards (which may act through local or
regional boards) have been designated as the water pollution control
agencies in their states to protect the water quality of the state.
Permit applications and proposed federal projects must be evaluated
for compliance with the applicable effluent limitations and the water
quality standards of the state. Certification of compliance with
these standards is required under Section 401 of the Clean Water Act,
and such certification must be obtained by federal or local projects.

State Coastal Commissions/Coastal Zone Management Agencies

Almost all coastal states have established agencies to administer
their coastal zone management plans in accordance with the federal
Coastal Zone Management Act. These plans and agencies differ from
state to state, but they are generally concerned with protecting,
maintaining, and enhancing the quality of the coastal zone, its
environment and resources; assuring orderly balanced use and
conservation of resources, taking into account social and economic
needs; maximizing public access to the coast and public recreational
opportunities; assuring priority for coastal-dependent/related
development, and encouraging initiative and cooperation in planning
and development in the coastal zone. As an example, the California
Coastal Commission, which is one of the most active state coastal
agencies, requires a master plan from each port and exercises appeal
authority to ensure that particular developments conform to those
plans. In addition, the California Coastal Commission issues coastal
development permits and can require that those permits include
mitigation measures if it is determined that substantial harm to
coastal resources will result.

In the case of Corps activities, no project is approved until
appropriate state agencies have confirmed that the proposed activities
comply with their coastal plans, or have waived their right to do so.

Advisory Council on Historic Preservation, State Offices
and Others (U.S. Army Corps of Engineers)

Under the National Historic Preservation Act of 1966 (16 U.S.C. 470a),
the Advisory Council on Historic Preservation was empowered to review
federal activities (including activities licensed by the federal
government) to ensure that listings in the National Register for

Historic Places will not be adversely affected. The Council published
regulations governing historic preservation (36 C.F.R. Part 800); the
Corps district ensures that they are followed, if applicable, as
specified by CEQ's NEPA regulations, which require taking into account
any significant aspects of environmental quality recognized by
federal, state, regional, or local governments, public entities, and
private organizations (such as the National Trust for Historic
Preservation). Most states have a historic preservation plan or
office. Among federal laws applicable to cultural resources are the
Antiquities Act of 1906 (16 U.S.C. 431), Archeological and Historic
Preservation Act (16 U.S.C. 469), Historic Sites of 1935 (16 U.S.C.
461), and Executive Order 11593 (May 15, 1971), "Protection and
Enhancement of the Cultural Environment" (36 Federal Register 8921).

THE PROCESS OF CONCURRENCE AND COORDINATION

Given the multiple responsibilities of many agencies for dredging
projects, it is little wonder that institutional decision making moves
slowly. The Corps, then, operates in a complex legal environment, but
it must also be emphasized that all dredging decisions and activities
exist in a political environment. In the political and legal
environment, it is understandable that the Corps seldom acts without
the concurrence of concerned agencies. This commitment to cooperation
and concurrence is so central to the Corps' operating procedures that
its regulations emphasize the point. For example, these regulations
state that the Corps will "...give full consideration..." to the
comments of the regional directors of the Fish and Wildlife Service
and the National Marine Fisheries Service" (33 C.F.R. 325.8(b)), "and
that the applicant will be urged to modify his proposal to eliminate
or mitigate any damage to...resources and in appropriate cases the
permit may be conditioned to accomplish this purpose" (33 C.F.R.
320.4(c)). The simple and compelling reality is that in the highly
pluralistic approval process for which the Corps is the coordinating
agency, consensus-building and concurrence among federal, state,
regional, and local agencies with major regulatory responsibilities is
now and will likely remain a fact of life. As a general rule, any
involved agency has a high probability of being able to block a given
dredging action or at least slow it down substantially if it is
strongly opposed to the action.

   A frequent problem that contributes to slowing down permit approval
is the lack of resources within the concurring agency to give the
permit careful review. Concurrence with Corps permits and Corps
projects is handled by the local and regional offices of concurring
agencies. Almost inevitably, local and regional offices face funding
and manpower constraints and must, therefore, set priorities for the
use of their limited resources. Understandably, the norm is for
concurring agencies to place highest priority on those programs for
which they have primary or lead responsibility. The effect of this is
that many agencies respond to requests for review of permits by
indicating that their resources preclude a response at this time but

that they reserve the right to respond later should they identify significant potential consequences in the areas of their responsibility or should they acquire additional funding and manpower resources. In sum, a non-review is not the equivalent of approval. Rather, it is a door left open through which a later modification or challenge may walk.

Problems attendant on reserved comment and multiple reviews affect local projects more than federal projects; first, because federal projects are scheduled for several studies, including a complete Environmental Impact Statement; but second (and perhaps more importantly), because the Corps district initiates the scoping process required by the NEPA guidelines of the Council on Environmental Quality (40 C.F.R. 1500-1508). In the scoping process, key officials from the other agencies are invited to help identify the principal items of environmental concern and how they will be addressed. This process is also affected by lack of manpower and travel funds, but the local Corps district will sometimes assist (Environmental Law Institute, 1981). During the time that elapses between the filing of the Draft Environmental Impact Statement and the final Feasibility Report, mitigation or other modifications may be considered, and consensus developed. Local projects await reaction from the review and comment period, usually, in contrast to the active process followed by the Corps for federal projects.

One of the most debilitating elements of the multiple agency involvement in dredging projects comes in the form of proposed mitigation actions. In some instances, mitigation actions are introduced late in the process and they require carrying out a redesign of the project. In other instances, differing agencies may recommend mitigation actions that are contradictory or contrary (Kenney, 1980; Dredging Committee of California, 1978; American Association of Port Authorities, 1981).

The coordination-consultation process followed by the Corps involves three phases: issuance of a public notice, a comment period, and a public interest review. In the case of public notices, the Corps is obligated by regulation to issue a public notice of its intent to issue a permit within 15 days of the receipt of a complete permit application containing all required information. The time between receipt of the initial application and the issuance of the public notice is often much longer. The most common difficulty is that the permit applications are incomplete--lacking required information, and the Corps must then request the additional information from the applicant. Not infrequently, applicants are slow responding to these requests for additional information. Many applicants lack the necessary expertise to know what is required by the permit or do not have the required information available.

Once the formal public notice has been issued, Corps regulations (33 Federal Register 94-31834) specify a 30-day comment period. The district engineer may grant an extension of up to an additional 15 days only if he determines it to be in the public interest. In fact, the Comptroller General of the United States (1980) has reported that Corps districts routinely grant single and multiple 15-day extensions,

88

most often at the request of other federal agencies, and that those requests for extensions are normally approved without the Corps' requiring the requesting agency to document why the time extension is necessary. If during the public comment period the Corps receives objections or significant substantive adverse comments, a period of negotiation is initiated. Particularly in the case of local projects, permits are not acted on until the proposers of the project and the objectors have worked out an agreement. The normal pattern is for proposals and counterproposals to ensue with delays in the negotiating process resulting from a lack of structure and management in the negotiating process (Comptroller General of the United States, 1980). Some have argued that this process could be significantly shortened if the Corps, with its expertise, were to participate actively in the negotiation and make its expertise available.

Once the public comment period is closed, the district engineer is required to make a public interest determination. In making this determination, the district engineer is supposed to consider the full range of interests by balancing the favorable and unfavorable consequences of a permit decision (47 Federal Register 31800).

The number of elements the district engineer is supposed to balance is truly impressive. He is supposed to carefully consider and weigh such factors as: conservation, economics, aesthetics, environmental concerns, wetlands, cultural values, fish and wildlife values, flood hazards, land use, navigation, recreation, water supply, water quality, energy needs, safety, food requirements, socioeconomic benefits, and the general welfare of the people. This public interest review is at once the heart of the Corps evaluation process and at the same time a process for which there exist no established criteria for weighing and balancing the various factors. One investigator (Rader, 1983) characterized the review process as follows: "The careful weighing of the benefits and detriments of the proposed activity called for in the regulations appear to have been replaced in practice by merely a display of those benefits/detriments. The jump in reasoning from a display of impacts to a decision implying a balancing of public interest factors is hidden. The public interest review has become an unaccountable informal process....The goal of balancing competing interests has been unfortunately ignored. This missing balancing effort appears to have been replaced with a policy that equates a lack of significant environmental impacts and/or unresolved objections from governmental agencies with a determination that the issuance of the permit would be in the public interest."

Although Rader criticizes the way in which public interest determinations are made, it is difficult to see how they could be handled differently. There are no professionally agreed upon, let alone politically acceptable mechanisms for balancing so many competing apples and oranges. The norm, then, appears to be that the Corps approves permits when there are no significant objections and either denies them or does not act on them when there are significant objections.

EACH PORT IS UNIQUE

The decision making system for regulating port dredging which has been characterized in the preceding reflects the general situation. That is, it is the result of national legislation, national regulations, and a general characterization of operating procedures. One additional critical point needs to be emphasized. That is, each port is unique. The common theme of the succeeding two chapters is that the design and construction of each port and the environmental protection actions required must reflect the specific characteristics of the individual port. Similarly, the political and economic circumstances and the complex of interest groups and governmental agencies involved also differ from port to port. Finally, in the case of federal projects, even at the congressional level, port authorization and appropriations and the specific parameters mandated for port construction are port-specific rather than programmatic. Any successful effort to bring speed, predictability, and stability to port dredging decision making must be based on the recognition that each port is unique.

Further, even if the decision making process can be speeded up and the time required to make decisions shortened, port projects still take years. During that time, both port needs and the understanding of the environmental and other implications of port dredging can change. For example, the movement toward larger bulk carriers and the use of containerships occurred rapidly. Similarly, understanding of appropriate handling of different dredged materials has increased rapidly in the last decade and offers resolution of many issues where previous uncertainty led to polarizations among participants. Any fast tracking system, then, needs to assure flexibility to allow consideration of the uniqueness of each port, and to allow efficient integration of new needs and understanding into the decision making process.

SUMMARY

The general pattern of decision making that characterizes the regulatory system governing port dredging is that decisions occur only when consensus is achieved. Consensus in this sense is defined as existing when no significant participants object so strongly to the action that they are willing to mobilize and oppose it. There is one other point that deserves emphasis and it is that in addition to all the inherent legal and political pressures for finding a consensus, the possibility of court action adds another. Particularly in the case of state and local agencies and private interest groups, strong objection to the approval of any permit or particular action within the context of federal projects can be opposed in the courts. Objectors do not necessarily have to win in the courts to win their point. If the courts provide a vehicle for substantial delay and that delay is costly to the proposers of the project, the threat of going to court becomes a powerful negotiating tool in the hands of

objectors. Although this point is not addressed formally in any of
the Corps or other agency documents, it is an essential reality for
all of the participants in regulatory decision making.

Few things struck the committee with more force than the frequency
with which participants in the dredging decision making system either
identified the courts as vehicles for slowing down the process or
mentioned concerns about possible court action. In sum, the
availability of the courts reinforces a widely noted characteristic of
the regulatory decision making process in the United States. That is,
that opponents and objectors deal within a system where the processes
are weighted in their favor. Simply stated, for anything to happen in
this regulatory decision making system, all significant participants
have to be in agreement at least to the point of not organizing in
opposition. Alternatively, to keep things from happening, only one
significant participant has to be vigorously opposed.

The Corps' lead role in the regulatory system for port dredging
reflects the need for consensus. While in specific instances it may
be possible to find the Corps acting over the objection of a major
participant, it is clearly not the norm. In fact, it is difficult to
conceive how the decision making system could be modified.

CONCLUSION

Many advocates of fast tracking have called for comprehensive
legislative regulatory change designed to streamline the decision
making system and accelerate decisions. Although these proposals have
much to recommend them in the abstract, they typically do not take
into account the need of the decision making system to balance an
extremely complex set of needs and interests. For this reason, some
of the proposals that have called for the Congress to concentrate
decision making in a single authority, presumably the Corps, such that
the Corps could act over the objections of major participants would
appear to be unrealistic. The committee, for example, could find no
instance of Congress having given any single agency such overweening
authority in circumstances similar to those existing for port
dredging. It is difficult to conceive of circumstances in which
Congress would free the Corps from the requirements of such
legislation as the National Environmental Policy Act, the Clean Water
Act, or the Endangered Species Act. Each of these pieces of
legislation was passed with broad public support; each has a strong
base of supporters in Congress; and each has a broad public
constituency. Survey data indicate continued broad public support for
environmental legislation. Congressional defenders of environmental
legislation appear to reflect the broadly held values of the American
people.

The preceding point was demonstrated by response to the Carter
Administration's proposal for establishment of an Energy Mobilization
Board. The proposed Board was to be given the power to fast-track
certain critical major energy projects. The proposal for the Board

was made at the height of the Iranian Energy Crisis, a time in which there was broad public sentiment for taking decisive action. It was widely believed that the inability to move rapidly on large energy projects resulted from circumstances very similar to those associated with dredging projects. Even in the crisis during which the Energy Mobilization Board was proposed, Congress refused to act. The coalition opposing an Energy Mobilization Board ranged from environmental organizations--concerned that environmental protection might be diluted--to political conservatives--fearful that state power and authority would be concentrated in a federal agency.

In the case of port dredging there is no similar crisis atmosphere and the possibility of building a majority to support a comprehensive legislative change making fast-tracking possible does not appear likely.

Alternatively, numerous specific proposals have been made for modifying particular regulations or particular coordination procedures (General Accounting Office, 1978, 1984; Comptroller General of the United States, 1980). The corps has already demonstrated the capacity to accelerate decision making in some of its own activities for federal dredging projects. Doubtless there are many opportunities for improvement and acceleration of decision making in this context. Optimism, however, must be qualified. Many regulations are the result of direct and specific mandates in legislation. For example, the need for a "no-jeopardy" biological opinion from the Fish and Wildlife Service is derived directly from the Endangered Species Act. In the face of that legislative mandate, no amount of regulatory change or organizational modification can overcome the requirement for Fish and Wildlife concurrence.

With regard to federal projects, the greatest opportunity for acceleration in decision making rests with Congress. With regard to its decision making, however, Congress is a law unto itself. Perhaps opportunities exist for accelerating decision making once Congress has found an answer to the funding issue. Short of that, given a funding stalemate, discussion of fast-tracking federal projects has an air of unreality.

WHAT ARE THE OPTIONS?

Any search for ways to accelerate decision making with regard to U.S. ports must start by recognizing certain facts. First, there is no formal statement of national port policy in the United States. Therefore, there are neither criteria nor a predetermined process for determining which of many competing ports should receive new or additional dredging, and in what order. Any major port dredging will result from one of two determinants: (1) ability of the individual port to convince Congress that its needs should receive first or high priority; (2) ability of the individual port to find and secure non-federal funding sources.

Second, there is a complex body of law and regulations which applies nationally to all ports.

Third, this body of law and regulations specifies generally applicable criteria which must be considered when making decisions about specific port activities. The laws and regulations direct that certain procedures be followed, and that for certain kinds of actions, specific approvals or permits be obtained.

Fourth, each port is unique. Responsibility for the development of the port is local. If the source of funding is federal, the port must mobilize and take those actions through its congressmen and senators to assure congressional authorizations and appropriations. Successful federal funding rests on the ability of the port (through its elected representatives) to mobilize the necessary majorities in Congress. Each port then is the locus of a micropolitical system for the congressional decision making process, and in the context of Congress, is in competition with other ports for congressional attention. Congressional decisions, then, require tradeoffs, compromises, and accommodations among the various ports.

Alternatively, if port development is locally funded, the ports must mobilize to assure that that funding is made available. Sources of funding for ports vary. Some ports are units of state government and funding requires actions by state legislatures. Others are entities of city government and require decisions within that context. And others are separate political entities which must determine within preexisting authorities how projects will be funded.

Simply stated, from the point of initial funding through every step in the process, ports in the United States are organized such that the major incentive and the major motive for action must come from the individual port. The U.S. port system is in some senses not a system at all: rather, a set of competing individual entities.

Ports are also unique in that they define their own needs. Some ports are predominantly bulk commodity ports; others, predominantly high-value cargo ports; and others, multicommodity ports. Based on the character of existing traffic and expectations about future potential, the needs and future capacities of ports vary.

Similarly, the physical environments of ports differ. Some of the ports this committee considers to be coastal are actually located on rivers or lakes some distance from the ocean; others that are coastal by any definition have very high or very low rates of sedimentation, comparatively, depending on littoral transport and protections; while still others (in fact, most ports in the United States) are in the complex sedimentary regimes of estuaries. Each port has a number of geographical, man-made, and other physical constraints: none has limitless physical possibilities for expansion or development.

Ports are also unique in the biological concerns of greatest importance: most coexist with (or near) other uses of the oceans and coasts, including commercial and sport fishing, public recreational areas, and productive wetlands; some ports are particularly concerned about toxic materials in their sediments; others must address a range of concerns pertinent to surrounding concentrations of population.

Each port is unique in the way it is organized. That is, ports may be state entities, local entities, independent political entities, and they are likely to have to deal with very different sets of interest groups. Some ports must deal with highly active environmental and social interest groups and others with few ongoing groups, although any port's plans might prompt citizen groups or action. Some ports are located in areas that have been the focus of major concerns by federal environmental agencies; and others in areas that have received little federal attention. Some have well organized, local commercial Chambers of Commerce and skillful leaders; others have less well-organized local commercial interests.

Finally, each port exists in a dynamic business environment. Its needs can change over short periods of time. Similarly, the character of environmental and social concerns can change rapidly.

Any effort to design a system to accelerate decision making for port development and to bring stability and predictability to the process, then, must begin with this recognition of uniqueness. From that follows a central conclusion: the achievement of predictability, stability, and speed will likely rest with the individual port, and any such achievement will require the development of procedures and processes that enable consensus to be achieved and sustained by interested participants. Stated differently, the possibility of creating a system where the lead federal agency, the Corps, has the capacity to accelerate port development over the objection of significant participants seems unlikely. Rather, fast-tracking requires creating conditions on a port-by-port basis in which the major organizational and interest group participants find themselves in sufficient agreement with proposed port developments so that they will not organize and mobilize to block or slow down developments.

Some years ago, a committee of the American Society of Civil Engineers recommended that the permit process for offshore and coastal development had to be rescued from the adversary process that the committee said immobilized it. That committee proposed that a consensus process should be strongly advocated by the ASCE (the organization declined, not being primarily concerned with policy matters). Such a consensus process will most likely evolve only with concerted direction and effort from the local port.

In general, conflict results from some combination of three conditions. First, participants in decisions differ on objectives or goals. Second, participants have differing understanding of facts. And third, some interested parties find themselves excluded from the decision making process. To minimize conflict arising from these conditions, each port needs to establish a planning process with a commitment to assuring that the planning process will be continuous. It is frequently said in corporations that the importance of planning is not so much the plan as it is the process. The same would appear to be true with regard to ports. The objective should be to develop a comprehensive plan for the port and ideally the port region, with the recognition that the planning process is the beginning of the consensus making process and that it is a continuous requirement. The starting point for any planning process must be to ensure that all interested parties are included. A key contribution of the planning process is the establishment and the maintenance of a communication system which keeps all interested parties informed.

Any planning process needs to identify the needs of the port both in the short and long term and the implications of those needs for the range of concerns reflected by the interested participants. The planning mechanism or process, then, needs to include all of the appropriate governmental agencies as well as port users, commercial interests, and environmental and public interests concerned about port development. What is required will obviously vary from port to port, but given the complexity of the issues that now surround port decision making and the unlikelihood of change at the federal level, any mechanisms other than those of local consensus-building appear to offer little chance of success.

REFERENCES

American Association of Port Authorities (1981), "Regulatory Review of Sec. 404 of the Clean Water Act and The Fish and Wildlife Coordination Act," Recommendations to Presidential Task Force on Regulatory Relief, Washington, D.C.

Comptroller General of the United States (1980), "Managerial Changes Needed to Speed Up Processing Permits for Dredging Projects," Report to the Chairman, Committee on Merchant Marine and Fisheries, U.S. House of Representatives.

Dredging Committee, California Marine Affairs and Navigation Committee, (1978), "A Muddle Over Mud," Presentation to Port Caucus, U.S. House of Representatives, July 13, 1978.

Environmental Law Institute (1981), NEPA in Action: Environmental Offices in Nineteen Federal Agencies (Washington, D.C.: Environmental Law Institute).

General Accounting Office (1984), Update on Army Corps of Engineers' Planning and Designing Time for Water Resources Projects (Washington, D.C.: Government Printing Office).

General Accounting Office (1978), Corps of Engineers Flood Control Projects Could Be Completed Faster Through Legislative and Managerial Changes (Washington, D.C.: Government Printing Office).

Kenny, M. (1980), "Port Permitting Problems," Coastal Zone '80 (New York: American Society of Civil Engineers), pp. 791-809.

Rader, C. D. (1983), "The Corps of Engineers Public Interest Review Process: Is It Working?" Coastal Zone '83 (New York: American Society of Civil Engineers), pp. 2086-2091.

8
Assessment of Technical Considerations and Needs to be Met in Dredging U.S. Ports

This chapter focuses on engineering design elements of navigational facilities, maintenance dredging, the capability of the dredging industry of the United States, and needs for research and development. Several technical considerations important to engineering design and dredging activities are treated in Chapter 7--estuary hydraulics, for example, and the site and nature of the disposal site for dredged materials. Another most important consideration--the institutional framework--is discussed in Chapter 7.

ENGINEERING DESIGN OBJECTIVES FOR DREDGED NAVIGATIONAL FACILITIES

There are two important design objectives for navigational facilities-accommodating the maneuvering requirements of vessels, and reducing as much as possible the future maintenance dredging required. Some general considerations of the vessel in the waterway and sedimentation are briefly described in succeeding subsections. It should be kept in mind that any engineered structure represents many compromises among these and other objectives, and for navigational facilities in particular, many unique local features have to be understood and taken into account.

Maneuvering Requirements of Vessels

Dramatic changes occur in a vessel's response characteristics in shallow water, and unique disturbing forces act on the vessel that have no counterpart in the open ocean. Vessels are primarily designed for the open ocean, however, so their accommodation in confined waters depends on adequate design of navigational channels and operational practices.

Entrance

For most ports there is a critical entrance (and exit) area seaward of the protecting headlands, rock, breakwaters, or jetties where both shallow-water effects and those of waves, swell, wind, and currents may act on the vessel. Where entrance channels are dredged in these transitional areas, greater depths and widths must be provided than those of channels within the port's sheltered areas, owing to the vessels' tendency to heave, pitch, roll, and drift (Marine Board, 1981; 1983).

Sinkage

Inside the entrance, shallow-water effects are accentuated by decreasing depths and widths of navigational channels. The velocity of water flowing around the sides and under the hull of the vessel must accelerate, with corresponding lowering of pressure (by Bernoulli's Law). The vessel sinks lower in the water with (usually) trim by the bow. For the same reasons, sinkage increases with the narrowness of the channel and with the vessel's forward speed. If underkeel clearance is small, vessel speed must be reduced to counteract sinkage, but it should be noted that minimum speeds must be maintained to counteract the forces acting on the vessel and to maintain headway. Some ships' engines (particularly the diesel engines favored in new ships) have minimum operating speeds. Sinkage is also affected by water density, and will increase in freshwater as compared to seawater (this is important to ports on river or estuarine systems, in which a change in water density will be experienced in a vessel transit).

Bank Effects

While water flow past the sides of the hull is symmetrical if the vessel is on the channel's centerline and aligned with it, moving off the centerline will decrease the flow area between the vessel and the near bank, causing the flow rate on that side to accelerate, with corresponding loss of pressure. This unequal pressure regime causes a bank-suction force aft, and a yaw moment turning the vessel back toward the centerline, as well as a sideslip velocity toward the near bank, that together with the vessel's forward velocity, induces a small drift angle toward the near bank; this also induces a small moment toward the opposite bank. Uncorrected, a vessel once off centerline would sheer from the near to the far bank, and back, or ground. Bank effects are forcing functions acting on a vessel that must constantly be corrected by steering changes.

Vessel Interactions

In passing or overtaking in navigational channels, vessels experience unique disturbing forces never experienced in the open ocean (the effects of vessel-vessel and vessel-bank interactions, since vessels must move off the centerline to pass or overtake). These effects develop fully after the vessels have passed, and in any area of passing or overtaking, sufficient width and length must be provided for some distance to allow controlled recovery.

Decreased Turning Performance

Vessels at sea have a turning radius comparable to their length, owing to the continuous sideslipping of water under the keel. This ability is lost in shallow water, particularly if underkeel clearance is very small, because water flow under the keel is constricted. Vessels in ballast also have decreased turning performance.

Winds and Currents

In winds or currents acting at an angle to the vessel, a compensating yaw (or "crab") angle must be achieved and maintained. This means that the vessel will "sweep out" a path broader than its beam. The very high superstructures of some vessels that have most of their profiles above water, such as containerships and car carriers, present considerable windage area, and may require more channel width than their narrow beams would suggest. Even vessels that have little profile above water fully loaded, such as tankers, may present considerably more windage area in ballast. The critical relationship appears to be the ratio of wind speed to ship speed: at ratios of wind speed/ship speed of about 6 to 7, great difficulty can be expected in controlling lightly loaded vessels or those with high windage areas, and at ratios of about 10, control of most fully loaded vessels will likely be impossible.

Prevailing winds blowing over long periods can also raise or lower water levels (wind setup or setdown).

Irregularities

A feature of navigation in channels and maneuvering areas that is often mentioned by pilots but that has not received much systematic study is the effect on vessels of bottom and bank irregularities. Modelling of navigational channels usually assumes uniform side slopes and unvarying bottoms, but general and local conditions usually favor rapid shoaling on one or another side of a bend or turn, or formation of a spit that encroaches on the channel at breakwaters or jetties, with the result of narrowing the width or reducing the depth of channels in locations where width and depth are most critical to maneuverability. Dand (1976) gives an example of ship collision caused by shoaling in a turn.

Piloting

Maneuvering a vessel in the shallow waters of a navigational channel and other port facilities is entirely different from maneuvering a vessel at sea in deep water with infrequent and distant traffic. Unlike a vessel at sea, a vessel in the confined waters of a navigational channel requires constant steering to counteract the number and magnitude of hydrodynamic forces acting on it. Harbor pilots familiar with the port and experienced in maneuvering vessels in confined waters board vessels and guide their transits in and out of the port. All the ports of the United States serving oceangoing traffic require pilotage.

Successful shiphandling by a pilot in navigational channels demands smooth, skillful integration of several very important elements: directing vessel movements; assessing other traffic movements in meeting and overtaking, as well as crossing traffic; evaluating waves and surges created by the ship; assuring that the helmsman clearly understands and executes rudder commands and steering directions without error; analyzing radar information; knowing the magnitude and effects of currents, wind, the hydrodynamic interaction of ship and channel; and anticipating possible changes in high-shoaling areas.

Harbor piloting in ports of the United States is typically of foreign vessels, of unknown maneuvering characteristics, designed and equipped primarily for the deep ocean. The pilot will therefore spend some time on boarding a vessel testing its responsiveness (and that of the helmsman), and checking the radar and other equipment. The ship's radar, in particular, might be in any state of repair or calibration. In poor visibility, the pilot must rely on the radar heading line, and a problem that frequently occurs with poor radar calibration is bearing resolution error. An undetected error in bearing resolution of 2°, for example, will place a vessel 200 ft out of position in just one mile. In some wind and sea conditions, and in heavy rain or snow, a "clutter" zone will appear on the radar screen representing the area around the ship. Activating the clutter-supression controls often eliminates small targets from the screen, such as buoys and fishing vessels. Losing buoys from the screen, the pilot may attempt to use the radar to determine the ship's position by estimating distances from prominent features of the landscape. An error in the ranging mechanism of just .05 mile will cause a position error of 300 ft.

Lack of Minimum Standards for Vessel Maneuverability

Considerably complicating the job of both pilot and channel designer is the lack of minimum standards for vessel maneuverability (Landsberg et al., 1983; Webster, 1983; Card et al., 1979). Even very modern vessels, and vessels in the same class, show wide variation in response characteristics, from relatively controllable to unwieldy. Recent efforts by the Society of Naval Architects and Marine Engineers (SNAME) and the International Maritime Organization (IMO) show promise of achieving such standards, but these efforts will take time.

Criteria for Dimensions of Dredged Navigational Facilities

General guidelines for the dimensions of dredged navigational facilities have been developed taking into account the over-all vessel maneuvering requirements described in preceding subsections. In the United States, the guidelines are developed by the U.S. Army Corps of Engineers (1983); consensus standards are also developed and updated by international organizations, such as the Permanent International Association of Navigation Congresses (PIANC), and the International Association of Ports and Harbors, and by other maritime nations. These standards (see Appendix B) are similar in most respects: those of the Corps tend to offer more guidance for smaller vessels, and those of international organizations to concentrate on large, full-form vessels, such as tankers.

The guidelines are based on selection of a design vessel or vessels, and calculating needed widths and depths for sinkage, passing or one-way traffic, wind and current effects, etc. The general guidelines offer a useful first approximation that must be refined with site-specific information and design validation.

The general criteria also provide standards for an initial assessment of existing facilities. Using the guidelines of PIANC and the Corps, the technical panel of the committee made a summary assessment of U.S. ports and a more detailed assessment of the navigational facilities of six ports, two on each coast, taking as the design vessels those that use the ports frequently. The results are briefly summarized in the succeeding subsection.

SUMMARY OF ASSESSMENT OF NAVIGATIONAL FACILITIES IN U.S. PORTS

Most navigational channels in the United States are made up of relatively short, straight sections between 1.5 and 1.7 nmi (nautical miles) in length, connected by turns and bends. A survey of all those with straight sections at least 30 ft deep (Atkins and Bertsche, 1981) indicates that the majority are less than 600 ft wide; the greater number of these being either between 350 ft and 400 ft or between 550 ft and 600 ft wide. More than 75 percent of the turns are 40° or less, 34 percent are between 20° and 40°, and 43 percent are 20° and less.

In comparison to the general criteria for navigational channels established by international organizations and the U.S. Army Corps of Engineers (1965, 1983), these dimensions are at or below the geometrical limits for the average-size vessels using the channels.

The technical panel of the committee found that ports in the United States generally lack adequate emergency anchorage areas, and that turning basins are few and minimal in dimensions for the vessels (not the largest) using the port. The question thus arose in the panel's investigation: what was the design basis--particularly the design vessel--for which existing navigational facilities were designed? Table 16 (Appendix G) shows the year of authorization for major navigational channels and turning basins at their present dimensions. Of 154 authorizations, only 34 have occurred in the past 20 years, 12 since 1970, and none since 1976. Some date from 19th century sailing ships.

Despite the paralysis in authorizations since 1976 (and some that were authorized in that and previous years have never been built), studies continue to be conducted of needed improvements (Table 17, Appendix G). All these proposed improvements were designed by the guidelines of the 1965 Engineer Manual, which predates the Corps' current 1983 Engineer Manual. While awaiting authorization (and as funds permit), updating occurs in the district offices by the new Engineer Manual. The Norfolk district, for example, indicates that in the interval awaiting authorization, studies have been undertaken to refine the design basis using the 1983 Engineer Manual, and alternative configurations for this project, as well as for the improvement of Mobile Harbor, have been tested using the full-scale vessel simulator (CAORF) of the U.S. Maritime Administration.

Many of the projects, however, represent minimal improvements for existing vessel traffic: the design basis assumes, for example, that design vessels will not be fully loaded, or width calculations are minimal, assuming tug escort. In general, many proposed improvements are for relatively modest sizes of vessels (which may or may not be appropriate), and not all proposals allow these vessels to be fully loaded.

It must be borne in mind that there are constraints on widths, depths, and diameters in many areas: existing berths, piers, and other structures; harbor and bay tunnels, bridges; submarine pipelines and cables; salinity locks, and water-supply intakes.

Nevertheless, the dredging projects have yet to be initiated to match shoreside improvements or the needs of vessels now calling regularly on ports of the United States. The principal engineering problem in the design of dredged facilities is time. As the proposed improvement progresses through successive stages of the process for gaining authorization and funding, the engineering refinement or redesign that might be undertaken is limited by the project dimensions established in the initial stages. Two proposed improvement projects now in progress through the decision making system have been succeeded by proposals for additional dimensions. In a previous study (Marine Board, 1983), the time and nature of the decision making process were found to discourage research and innovation, and to impose limits on engineering, owing to the long times that elapse between the initial assessments of need and the initiation of dredging. More importantly, the time scale of the process was found to exceed the time scale of major changes in the world fleet.

Thus, while the general criteria for the design of navigational facilities have been brought up to date, institutional issues have impeded their effective application.

Operational Adequacy

The time and nature of the institutional process for achieving improvements in navigational facilities, and the funding stalemate of the past 10 years imply increasing obsolescence of the ports' waterways. Significant obstacles prevent the systematic application of engineering and construction dredging to ensure navigational adequacy. The burden to achieve navigational adequacy then falls on operations--on the conditions and practices used in individual ports.

To gain an understanding of the operators' views of navigational facilities in U.S. ports, the technical panel sent a questionnaire to the pilots organizations (Appendix C) requesting information about channel size and design, maneuvering problems, aids to navigation, maintenance dredging, and operational strategies used, if any, to compensate for perceived physical inadequacies. Of the organizations responding, only 2 judged the channels adequate for present vessel traffic; 3 suggested that channels would be adequate if maintenance dredging were performed on a regular basis; and 7 indicated that the channels were inadequate.

Vessels named by the pilots as being most difficult to handle divide into two groups (some organizations mentioned both): the largest, deepest-draft vessels they handle, owing to small underkeel clearance, and lightly loaded vessels with high, flat sides, such as containerships and car carriers (as well as a passenger vessel, in one case, having a high abovewater profile). Among the areas in their ports pilots most frequently cited as critical were jettied entrances, followed by narrow sections and tight turns. Other critical areas mentioned were those where crosscurrents or crosswinds are encountered. One pilot group in the Pacific said their entrance channels were adequate in normal conditions, but inadequate in swells. The pilots were unanimous in the judgment that improved aids to navigation cannot substitute for channel improvements.

All respondents indicated that special operating arrangements have been established by the pilots organizations to compensate for inadequate channel dimensions: one-way traffic, restricted passing and overtaking in bends and turns, transit with high tide for underkeel clearance, and use of tugs. In certain channels, pilots use hydrodynamic interactions with banks and other vessels to execute meeting and passing situations, or to round a turn of inadequate radius of curvature (using sheering effect to augment the decreased turning performance of a large vessel with small underkeel clearance).

It is important to understand these operational practices in the design or improvement of navigational channels; for example, observation of critical maneuvers often shows less variation in swept paths among pilots than in less-critical maneuvers, but this may indicate an area that needs widening, rather than one that could be

narrower. Control of a vessel in critical maneuvers is often achieved
by a great many rudder commands and a higher average value of rudder
angle (Hooft et al., 1978).

In a review of channel design, Hooft (1981) recommends a
sensitivity analysis of the vessel's controllability as a function of
external factors (such as wind or current) and channel width. Where
two-way traffic is frequent and widening is indicated but not
possible, it is helpful to have emergency anchorages alongside the
channel.

It should be noted that all calculations or estimates having to do
with the navigational requirements of vessels will be accompanied by
some uncertainty:

- The behavior of vessels in channels (although better understood
  today than in the recent past) is still very much in need of
  further study. Little exact guidance is available, and actual
  behavior may differ from predicted behavior owing to a number
  of complex and interactive factors.
- Computer-aided vessel simulation has improved in recent years,
  offering the potential for engineering design verification of
  alternative dimensions and layouts. Caution must be exercised
  against excessive fineness in the determination of channel
  dimensions through vessel-transit simulation, as even the most
  sophisticated simulator is accurate only within about a 20
  percent range.
- Local conditions of the physical environment are important but
  highly variable. The ship's response, in turn, is affected by
  its velocity, hull configuration, propulsive mechanism,
  loading, and underkeel clearance.
- There are no minimum standards for vessel maneuverability.
- Even more importantly, there are no consensus standards for
  navigational safety. This was identified as a top priority for
  the design of entrances to ports and harbors by an
  interdisciplinary meeting (Marine Board, 1981). Some
  shipowners have developed probabilistic methods to enable their
  ship's masters to calculate underkeel clearance and thus
  determine the advisability of entering ports around the world
  (Kimon, 1982). This method is data dependent, and can be
  improved with more and better data.

DESIGN OF NEW CONSTRUCTION DREDGING
PROJECTS FOR MINIMAL MAINTENANCE DREDGING

As pointed out in Chapter 9, thorough understanding of local tidal
hydraulics and circulation is necessary to design dredged navigational
facilities for minimal shoaling (see also Marine Board, 1983).
Site-specific hydrographic surveys, measurements of currents, and an
understanding of existing patterns of sedimentation in the port are
all necessary; in addition, a physical model can be a helpful tool in
assessing interactions of the facility with currents and (possibly)

effects of the facility on salinity distributions. If an evaluation
of the effectiveness of the design is needed, models of the currents,
the salinities, and the sediment transport will be required. The
Corps has conducted research, development, and field studies for many
years to improve its ability to model the hydrodynamics, salinities,
and sediment transport in waterways, and the private sector also has
the capability to make measurements and provide physical and
mathematical modeling services. Many of the processes of aggregation,
deposition, and erosion important to an understanding of sediment
transport, and thus, the management of sediment deposition have been
incorporated in mathematical descriptions for quantitative evaluation
of design and management alternatives (Ariathurai and Krone, 1976;
McAnally, 1984). The perpetual nature of maintenance dredging argues
for investment in site studies and models to guide design and
subsequent management.

New construction dredging projects offer the opportunity to reduce
subsequent maintenance dredging by design and management strategies.
In this connection, it might be noted that in many ports, federal
projects and local projects (particularly side channels), together
with the location and orientation of piers, wharves, and other
pile-supported structures, are incompatible. That is, one causes
accelerated shoaling for the other. A coordinated plan would be
helpful in reducing these incompatibilities and reducing maintenance
dredging.

A more difficult conflict is that between the need for emergency
anchorages and the disproportionate amount of maintenance dredging
these facilities typically require. The same is true of turning
basins, but their economic yield in terms of accommodating vessels is
perhaps more evident. Little engineering attention has been given to
emergency anchorages and turning basins. One possible solution for
some ports would be to dredge the facilities with flatter, stepped
side-slopes. The design would have a higher initial cost, but far
lower maintenance cost. Another interesting possibility is being
investigated by the Norfolk District of the Corps: using anchor buoys
similar to those developed for offshore oil loading/unloading
(described in Chapter 5) for offshore anchoring.

MAINTENANCE DREDGING

Table 18 (Appendix G) shows the annual average maintenance dredging
costs for each port. As the total approaches a half-billion dollars a
year, reducing the sedimentation associated with navigational
facilities, and achieving the lowest-cost maintenance dredging program
are important dredging needs.

Determining the most cost-effective program of maintenance dredging
depends on detailed site-specific knowledge (Herbich et al., 1981;
Marine Board, 1983). As indicated in Chapter 9, navigational
facilities change the preexisting sediment regime; therefore, an
important consideration in reducing maintenance dredging requirements
is the siting and design of these facilities. Other considerations,

such as improvements in dredging plant and its use, are discussed in succeeding sections.

Rates of deposition and types of sediments vary greatly from port to port, and man's activities near and far from the port, as well as natural causes, make significant contributions that cannot always be predicted or controlled. In some areas, most of the annual sediment movement will occur during a few storms. Waves and surges generated by the vessels can also move sediments; over time, bank erosion from these forces can modify the channel's side slopes (Herbich and Schiller, 1984). As a result of these and other in-channel forces, the channel ages and changes shape, with corresponding shifts in areas and rates of sedimentation.

Thus, determining an effective maintenance dredging program in a particular port depends on a great deal of historical and current local knowledge, and frequent hydrographic surveys. The usual case for an existing navigational facility is that some areas have higher shoaling rates than others, and deciding when and how much additional dredging they should have also depends on frequent hydrographic surveys. Trawle and Boyd (1978) found hydrographic surveys to be infrequent in the Corps districts, and substantial variation among the districts in the methods used to calculate the amount of additional dredging needed in these areas. Since the 1978 report, the Corps has made considerable investments in vessels and survey equipment for the districts. The information collected by the committee and technical panel indicates that survey practices and advance maintenance dredging (deeper dredging in selected areas) still vary from district to district.

One impediment to more efficient maintenance dredging (discussed in a succeeding section) is the year-to-year budget of the Corps. As funding for operations and maintenance has declined in constant dollars, the Corps has distributed the gap among the districts. In reading the yearly reports of Corps activities, it can be seen that the major projects not being maintained at project depth change from year to year, as some will be dredged and others allowed longer times between maintenance dredging.

An important set of considerations that is sometimes not addressed by maintenance dredging programs is that the most efficient operation of the port depends on assured access by vessels at the drafts specified in port guidelines. Port calls by liner operators, in particular, are scheduled months in advance. Many ports allow transit of deeper-draft vessels at high water, or in one-way traffic, or some other combination of operational practices to ensure passage at small underkeel clearance. These smaller underkeel clearances--about 2.5 percent of vessel draft--mean that the vessels are transiting at closer tolerances than those for which the channel was designed, and maintenance dredging is more critical. Even if vessels avoid grounding in areas of higher deposition, the presence of shoaled areas can affect their response characteristics, and this can be equally critical in narrow channels at small underkeel clearance.

Assessment of the adequacy of maintenance dredging in the ports of the United States would entail detailed port-by-port analysis and site studies, including a review of historical data on dredged volumes and frequencies, and of the changes over time in the facilities and their use. The committee and technical panel gained the general impression that maintenance dredging was a high priority in some districts, characterized by frequent surveys, long-term planning, and advance maintenance dredging, and a lower priority in others, characterized by a more reactive program of responding to the needs expressed by representatives of the port, pilots, or local U.S. Coast Guard.

CAPABILITY OF THE DREDGING INDUSTRY

A survey for the International Association of Dredging Contractors (Prognos, 1984) indicates that every developed maritime nation funds the new construction and maintenance dredging of its major navigational facilities, and that every nation is concerned to keep down the costs. The report recommends that (where appropriate) dredging be contracted to the private sector, a solution that has already been instituted for the most part in the United States. Several questions have been asked about the dredging industry in the United States: If a significant number of new construction dredging projects were initiated, would the industry have the capability to perform the work? What can be done to lower the cost of dredging? What technical improvements can be made for greater efficiency and productivity? These questions are taken up in the following sections.

Equipment and Procedures

The dredging of sedimentary deposits within ports and navigational waterways is accomplished by one of two primary techniques, hydraulic or mechanical. Within each class, a number of functionally different systems are available (see Figure). The ultimate selection of the operating system is based primarily on the sediment type, water depth, sea conditions, location and proximity of the disposal area, and to

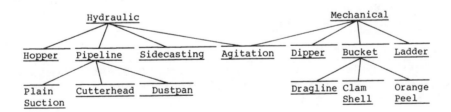

Dredging Systems

some extent, the availability of equipment. In addition, the contamination levels of the sediment and the need to minimize near-field resuspension and far-field dispersion may be considered (as indicated in Chapter 9).

The majority of dredging projects in the United States employ hydraulic dredging techniques (Table 19, Appendix G). These techniques are particularly well suited for use in areas characterized by a high degree of sediment mobility where virtually continuous dredging is required and the dredged material is either moved from the channels to disposal areas in deeper water or placed in reasonably proximate shoreside containment areas. Mechanical techniques are more frequently employed in areas of slower sedimentation. These techniques also appear to be favored if coarse-grained material is to be dredged, or if high contaminant levels require minimal agitation or fluidization of the sediments and a general retention of the cohesive character of in-place, fine-grained materials. These latter characteristics, in combination with the limited number of alongshore disposal areas, have historically favored the use of mechanical dredging techniques in New England.

Structure of the U.S. Dredging Industry

The U.S. dredging industry consists of approximately 190 firms* competing primarily for federal contracts. The ten larger companies account for 56 percent of dredging under federal contracts.

A recent study by the Small Business Administration concludes that federal procurements account for about 75 percent of all dredging in the United States. Given average annual federal contracting of $331 million for the period 1980-83, the industry performs about $440 million of work annually. Additionally, the Corps of Engineers operates a fleet of 13 dredges which performed an average of $86 million per year for the same period. The following table summarizes dredging revenues in the U.S.

Annual Average Value of Dredging Work in U.S. (1980-1983)
(millions of dollars)

	Contractor	Corps of Engineers	Total
Federal contracts	$331	$86	$417
Private contracts	110	0	110
	$441	$86	$527

SOURCES: Federal contract dollars from U.S. Army Corps of Engineers. Private contract dollars from Small Business Administration, 1984.

---

*Small Business Administration (1984) estimates 250, but without evidence. A total of 163 bid successfully on federal contracts from (1980-1984); 31 more bid unsuccessfully on at least one project.

The distribution of cubic yards dredged in federal projects from 1980 through 1983 is shown below.

Total Amount of Dredging (10^6 yds^3) (Federal Projects), 1980–1983

Dredged	1980	1981	1982	1983	Four-Year Total	Average
By contractors	225	281	220	250	976	244
By USCE	84	88	60	50	282	70
Total	309	369	280	300	1,258	314

SOURCE: U.S. Army Corps of Engineers

Data on cubic yards dredged under private contracts are not readily available. If they were roughly proportional to the average price per cubic yard of federal contracts, they would not exceed 75 million cubic yards annually. In all likelihood, the true figure is much lower because most private contracts are for relatively small quantities with correspondingly higher unit costs than federal contracts.

Having applied average price per cubic yard to work performed under federal contracts to estimate work in the private market, a note of caution needs to be added about making dollars-per-yard comparisons between the contractor fleet and the Corps fleet.

Comparisons using annual averages or totals are virtually meaningless owing to differences in types of projects, measurement of yards dredged, and equipment utilization rates. Contractor dredges perform virtually all cutter and bucket work, half the hopper work, and about one-fourth of the dustpan work, and Corps dredges perform the balance. Cutter jobs often involve sizable preparation of disposal areas that account for 10 to 20 percent of contract price while material dredged by bucket, hopper, and dustpan dredges is usually transported to a deep-water disposal site. Contractors normally work on unit-price contracts and are most often paid on the basis of quantities determined by before-dredging and after-dredging surveys. Corps dredges, on the other hand, work until surveys or the depths of operations show that desired depths and widths have been achieved. The Corps is less concerned about overdredging, which is uneconomic for a contractor. Contractor dredge production is usually measured by net pay yardage while Corps dredge production is measured by gross yards removed. Finally, the Corps fleet has about a 70 percent utilization rate while contractor dredges average less than 50 percent utilization and thus must spread their fixed costs such as depreciation, insurance, and interest over proportionately fewer yards.

In a presentation to the Dredging Committee of the American Association of Port Authorities in 1981, two industry representatives described the U.S. dredging fleet and its annual production capacity in millions of cubic yards as follows:

	Number of Dredges	Annual Capacity
Large Cutter Dredges (18" to 42" Diameter Discharge)	101	453
Small Cutter Dredges (Discharge < 18")	150	
Large Bucket Dredges (12 to 22 c.y.)	18	164
Small Bucket Dredges (5 to 10 c.y.)	60	
Hopper Dredges (1,200 to 12,00 c.y.)	11	83
Dustpan (38" Discharge)	1	11
	341	711

The most sigificant change to the contractor fleet since then has been the addition of two more hopper dredges (4,000 and 2,800 cubic-yard capacity, respectively) and the keel-laying for a third hopper dredge of about 4,000 cubic-yard capacity. Since few dredges have retired or left the country to work overseas, industry capacity has remained in the neighborhood of 700 million cubic yards per year. Based on the peak workload of 281 million yards dredged by contractor plant for the federal government in 1981, utilization stands at about 40 percent of physical capacity. Thus, the industry has substantial extra capacity available for private work and for new work dredging.

The Corps of Engineers' fleet consists of 13 dredges:

Large Cutter Dredges (>18" Discharge)	2
Hopper Dredges	4
Dustpan Dredges	3
Sidecaster Dredges	3
Special Purpose Dredge	1
Total	13

One cutter dredge is scheduled for retirement during fiscal year 1985. The total does not include a number of small two-man cutter dredges which have very low utilization. As recently as 1980, the Corps fleet consisted of 27 active dredges. The Corps has retired dredges as contractors have built new plant under the terms of the Industry Capability Program discussed in more detail in a succeeding subsection. The current fleet of 13 dredges includes 3 hopper dredges launched in 1982 and 1983. The average annual workload of $86 million gives the Corps about 16 percent of the U.S. dredging market. Only one contractor performs a larger share of total U.S. dredging work than the Corps of Engineers' fleet.

Improving the Economy and Efficiency of Dredging

Greater economy and efficiency in dredging can be achieved by replacement of plant with modern dredges, application of available technology (instrumentation, automation), and integration of project planning. These are briefly discussed in succeeding sections.

Replacement of Dredging Plant

Most of the dredging in the United States is performed by cutter-suction dredges, with hopper dredges claiming the next-highest percentage, and dustpan, clamshell, and dipper dredges the remainder. The dredging industry in the United States has invested substantial sums in recent years to replace the entire hopper dredge fleet with modern, technologically efficient dredges. Therefore, the most effective improvement in over-all dredging efficiencies can be realized in the modernizing the cutter-suction fleet.

Cutter-Suction Dredges: Problems and Opportunities

The cutter-suction dredges of the United States are relatively old. Only 5 of the 20 largest were built in the last 10 years. This fleet, therefore, lacks most of the technology developed in the last decade. Another characteristic of cutter-suction dredges that contributes to inefficiency is their general-purpose nature. They were usually designed to handle the "typical" project rather than to have the optimum capabilities for a specific project. They are normally too powerful for the simpler projects or too weak for the more difficult jobs.

The benefits of replacing this equipment are many. Available technology increases productivity at reduced operating cost, and design features can be added that expand capabilities and enhance safety.

The high capital cost of this plant is the major impediment to replacement. Attracting the necessary capital to build the new dredges will require changes in the market for which they compete (as described in a succeeding section).

Among the new equipment for cutter-suction dredges are the dredging wheel and suction tube position indicator system.

Dredging Wheel The dredging wheel replaces the cutter on a cutter dredge. In the dredging wheel, the buckets are bottomless. By placing the buckets close together and overlapping, a tunnel is created, the inner limitation of which is the suction mouth itself. The dredging sequence is mechanical excavation followed by hydraulic suction.

Position Indicator System This type of indicator provides the operator with immediate visual indication of the position of the suction pipe, depth of the suction head and the angle of the lower part of the suction pipe in both the horizontal and vertical planes.

Recent Improvements in Hopper Dredges

Although no single type of dredge will ever be universally superior to all others, the hopper dredge is the only general-purpose plant that can work effectively in open water subject to the action of waves.

The three most important parts of a trailing suction hopper dredge are the hopper, the suction draghead, and the dredge pumps. Recent improvements have been made to these parts (Herbich and Brahme, 1980).

Hopper Turbulence in the hopper maintains the dredged material in suspension: to allow the material to settle quickly, it is important to keep the turbulence to a minimum. Recent developments (Brahme and Herbich, 1977) include installation of the discharge pipes farther down into the water at mid-depth, or even below, and discharging sideways at the aft end of the hopper. Gratings have also been provided on two sides to reduce turbulence.

Draghead-Mounted Dredge Pump One of the significant improvements in recent years is installation of a dredge pump on the draghead. As a result, the suction pipe has become a delivery pipe. It was possible to achieve a specific gravity of 1.4 in the solids-water mixture, even when the dredging depth was increased.

Active Draghead This new type of draghead was developed to achieve economically acceptable output from a hopper dredge operating in clay. The draghead is called the "active rotary draghead." It incorporates a rotating cylinder with a number of knives that slice the clay layers.

Venturi Draghead The Venturi draghead consists of three parts the pivoting part, called a visor, the fixed part, which contains the water jets, and an elbow transition between the fixed part and the actual suction tube.

The operating principle of the Venturi draghead is based on creating negative pressure immediately above the seabed by converting part of the pressure energy into kinetic energy. It appears from the field tests that the production in fine sand can be increased by 30 to 40 percent. However, no increase in production was observed in dredging of coarse sand.

Automatic Draghead with Winch Control System The automatic draghead winch controller was developed to regulate the movements of the suction pipe and draghead throughout the dredging cycle. It is programmed to swing the pipe outboard, to lower the pipe, and, in conjunction with the swell compensator, maintain the correct pressure of the draghead on the bottom.

The installation of an automatic suction pipe controller has simplified the operational procedures, thus enabling the operator to concentrate on obtaining the maximum output of solids. It has also enabled the vessel to continue operations in bad weather, while minimizing the risk of damage.

Split-Hull Hopper Dredge A hopper dredge divided longitudinally into two parts, which are joined by hinges at the main deck level, is emptied by allowing the two halves to swing apart. The main advantage of this type of dredge is the fast disposal of material and easy disposal of sticky clays, clay loam, and silt.

Underwater Pump in the Suction Pipe A pump in the suction pipe supported by a ladder not only increases the dredging depth but also increases the efficiency of the dredging process (Herbich, 1975).

Automation

The chip-based microcomputer technology opened the way for automation in the dredging industry. Automation assists the operator, but does not replace him. By taking over data acquisition, providing real-time data analysis and displaying operations of the various elements, the operator will be able to follow the process more carefully, and will be able to take steps to improve the efficiency of the dredging project.

Individual automatic systems that have already been developed are vacuum-relief valve, bypass valve, automatic draghead winch controller, automatic light mixture overboard, and draghead visor controller (Van Zutphen, 1983). For example, the automatic suction pipe controller moves the suction pipe. The controller actuates the winches to swing the suction pipe outboard or inboard and alters its position during dredging. It also controls the swell compensator and incorporates a number of safety systems.

Production Instrumentation

Production Meter

A production meter system provides continuous indication of density, total flow, and solids mass-flow rate of the material pumped by a hydraulic suction dredge (Figure 4, Appendix G). A production meter system can also give total solids production (Erb, 1981).

Two types of density gauges are commercially available: a differential-pressure gauge and a nucleonic density gauge (Figure 15). The magnetic flow meter measures the total flow rate. Other types of meters, such as the sonic flowmeter and Doppler flowmeter, are under development, or have recently become available (Roskam et al., 1983).

As shown in Figure 5 (Appendix G), the leverman may operate as part of a feedback loop. The leverman monitors the operation of the dredge by watching the indicator and adjusting the controls as necessary. Usually three parameters are displayed to the leverman:

- slurry density,
- flow velocity, and
- solids flow rate.

The information is best displayed by use of a crossed pointer display of the type shown in Figure 6 (Appendix G).

Nuclear Silt Density Meter   A nuclear silt density meter (Belgraver, 1983) has been used to measure the material in situ in connection with the "nautical depth" concept (Marine Board, 1983).

Economies in Efficiency

Dredging operations in Europoort-Rotterdam were significantly improved by the introduction of modern partly automated dredges and the installation of modern instrumentation. The costs of annual maintenance dredging were reduced 40 percent in spite of inflation and fuel-price increases. This is a good indication that significant economies can be achieved by modernization of the equipment used.

Market Incentives and Effects

Market expansion, such as the port deepening projects being considered in the United States, would likely stimulate investment. Caution must be expressed, however, that this type of sudden expansion will also result in higher prices for dredging in the short term. The longer-term result would probably be an improved fleet capable of producing more efficiently. Capacity will increase with demand and force prices to moderate. This pattern is suggested by examining the international dredging market in the mid-1970s. There was an unprecedented rise in demand owing to the port development programs in the Middle East. This forced prices up and encouraged investment in new plant and equipment.

Three years later, a new fleet of dredges was available. This increase in supply and a moderating market combined to bring the price level well below that which had prevailed. The users of dredging services paid a premium for several years but now are enjoying savings generated by a more efficient fleet.

Changing market conditions for hopper dredging in the United States produced a marked difference in the 1970s. Until the late 1970s, hopper dredge work was performed by a fleet owned entirely by the government. In 1979, legislation was enacted to allow the private sector to compete for a majority of this work. This was a deliberate decision to promote private-sector capability by sacrificing short-term savings from lower-cost (in depreciation and interest), but obsolete dredges. In 5 years, the dredging industry invested more than $250 million in efficient, productive hopper dredges.

Integrated Projects

With few exceptions, dredging work in the United States is carried out under short-term contracts. The majority of dredging work is funded and administered by the U.S. Army Corps of Engineers, and for a number of economic and regulatory reasons, the projects are segmented. A typical contract is for about 4 months of work, for $2 to $3 million.

The proliferation of smaller projects prompts use of older, less-sophisticated equipment. Although dredging costs are kept low over the near term, this practice discourages investments in new dredging plant and equipment, and in research and development.

There are several specific costs associated with small segmented projects:

- Mobilization--This category includes all costs associated with transporting equipment, people and materials to and from the site. They also include setting up or rigging the dredges for operation, establishing supply lines and complying with the lengthy administrative procedures associated with each contract. These costs often account for 15 to 20 percent of the costs on small contracts.

- Learning Curve--Each project is different. Often crews are unfamiliar with the idiosyncrasies of a project and must gain experience with the project before the dredge output reaches its maximum. This cost can be quantified. In comparing average production rates, it is often found that production is as much as 50 percent higher during the second half of a project than it is during the first. Additionally, costs per unit of time are lower during the later stages of most projects.

- Advance Maintenance--Additional material can be excavated by dredging somewhat deeper without decreasing the forward progress of the dredge. This deepening can be added at little extra cost, and will increase the interval between dredging.

CONCLUSIONS

Although general criteria have been developed for the design of dredged navigational facilities, their application in the United States is impeded by the length and character of the decision making process.

One of the consequences of the long lead times for decisions about port dredging is to discourage systematic engineering for port development. The concept of the design vessel in studies of dredging projects is hardly applicable to a world fleet that has a half-life of 10 years when the approval process takes more than 20 years. None of the existing authorizations for dredged navigational channels is as recent as the advent of large dry-bulk carriers or the latest-generation containership. The lack of timely improvements places the burden on local pilots, the ports, and U.S. Coast Guard (but primarily the pilots) to develop operational practices that enable vessels to transit obsolete navigational facilities safely. Besides the reduction of safety margins that would otherwise be achieved through engineering design and maintenance dredging, these practices are uneconomic in many ports.

The institutional process is also project-specific rather than programmatic. A programmatic approach is needed to achieve optimal port design to minimize maintenance dredging as well as port efficiency and navigational safety.

Each port is unique; thus, site studies are most important in defining dredging problems. Much remains to be learned about the maneuvering requirements of vessels, and collaborative interdisciplinary efforts are needed to achieve better understanding and to refine the tools of design verification and analysis (such as vessel simulation). Mathematical and physical models, field measurements, and engineering observation are well developed and need to be employed to understand local sedimentation and to minimize the maintenance dredging required by new construction projects.

The dimensions of dredged navigational facilities in the U.S. appear minimal for the vessels now using them, and emergency anchorages are small or lacking. Depending on vessel traffic and other local conditions, these minimal dimensions may be adequate, but the institutional process for improvements will impede needed improvements if they are inadequate. The process is insufficiently flexible to allow timely spot improvements, such as widening a turn.

The committee did not attempt a thorough evaluation of the nation's maintenance dredging program, as this would have entailed very detailed port-by-port analysis. The committee notes, however, that the year-to-year budget of the Corps and its declining level in constant dollars is an impediment to the efficiency of the maintenance dredging program and to the operations of the ports. That is, the Corps attempts to achieve equity among the ports by lengthening the time between dredging intervals at successive ports, and those suffering the lack of authorized depths must restrict ship drafts during those periods. Greater use might be made of advance maintenance dredging in high-shoaling areas, and to widen turns.

Existing institutional arrangements also restrict the Corps from making the most effective use of dredging resources.

Most of the dredging in U.S. ports is carried out by the private sector under short-term, unit-price contracts to the Corps (and a smaller amount under local port contracts or by dredges owned by the port). Greater economy and efficiency in dredging can be achieved by the replacement of existing plant with modern dredges, application of available technology in instrumentation and automation, and integration of project planning. The U.S. fleet of hopper dredges is modern and technologically efficient: the principal opportunities for improvement are in the cutter-suction fleet. The proliferation of small dredging projects (owing to economic and regulatory constraints) prompts the use of older, less-sophisticated equipment, and additional costs for repeated mobilization, inability to maximize the productivity associated with the later stages of a larger project, and loss of the opportunity to perform advance maintenance dredging at small additional cost. Positive changes will be needed to provide the market incentives for investment in new dredging plant, and changes in institutional arrangements will be needed for other improvements to be made.

115

REFERENCES

Ariathurai, C. R. and R. B. Krone (1976), "Finite Element Model of Cohesive Sediment Transport, "J. Hydr. Div., ASCE, HY3: 323-338.

Atkins, D. A. and W. B. Bertsche (1981), "Evaluation of the Safety of Ship Navigation in Harbors," Problems and Opportunities in the Design of Entrances to Ports and Harbors (Washington, D.C.: National Academy Press).

Belgraver, N. J. (1983), "A Nuclear Silt Density Meter," Dredging Engineering Short Course Notes, Texas A&M University, College Station, Texas.

Brahme, S. B. and J. B. Herbich (1977), "Dredging in India, Suggested Improvements in Techniques and Equipment," Report No. CDS 204, Texas Engineering Experiment Station, College Station, Texas.

Card, J. C. et al. (1979), "Report to the President on an Evaluation of Services and Techniques to Improve Maneuvering and Stopping Abilities of Large Tank Vessels," Report No. CG-M-4-79, Washington, D.C., U.S. Coast Guard.

Dand, I. (1976), "Hydrodynamic Aspects of Shallow Water Collisions," Naval Architect, 6: 323-346.

Englehardt, E. H. L. (1983), "Draghead Position Indicating System," Dredging and Port Construction, February.

Erb, T. L. (1981), "Production Meter Systems for Suction Dredges," Dredging Engineering Short Course, Texas A&M University, College Station, Texas.

Heiberg, E. III (1983), "Recent Port Planning Developments in the United States: Status of Channel Deepening Proposals for Major United States Harbors," 8th International Harbour Congress, 2.109.

Herbich, J. B. (1975), Coastal and Deep Ocean Dredging (Houston, Texas : Gulf Publishing Co.).

Herbich, J. B. and R. E. Schiller (1984), "Surges and Waves Generated by Ships in a Constricted Channel," Presentation to 19th International Conference on Coastal Engineering, Houston, Texas, September 1984 (proceedings in press).

Herbich, J. B. and S. Brahme (1980), "Modern Developments in Dredging Equipment," Dredging Engineering Short Course, Texas A&M University, College Station, Texas.

Herbich, J. B., W. R. Murden, and C. C. Cable (1981) "Factors in the Determination of Cost-Effective Dredging Cycle," Proc. XXV Int. Nau. Cong., Inland and Maritime Waterways and Ports, PIANC, Section II, Vol. 2, Edinburgh, Scotland, May 10-16.

Hooft, J. P. (1981), "Ship Controllability," Problems and Opportunities in the Design of Entrances to Ports and Harbors (Washington, D.C.: National Academy Press).

Hooft J. P. et al. (1978), "Horizontal Dimensioning of Shipping Channels," Presentation to STAR Symposium, Society of Naval Architects and Marine Engineers, New London, Conn., April 1978.

Landsburg, A. C. et al. (1983), "Design and Verification for Adequate Ship Maneuverability," Presentation to Annual Meeting, Society of Naval Architects and Marine Engineers, New York, November 9-12, 1983.

Marine Board (1983), <u>Criteria for the Depths of Dredged Navigational Channels</u> (Washington, D.C.: National Academy Press).

McAnally, W. H., Jr. (1984), <u>Hydraulic, Salinity, and Sediment Models,</u> Report No. TABS2, Waterways Experiment Station, U.S. Army Corps of Engineers, Vicksburg, Miss..

Prognos, AG, (1984), <u>A New Policy to Cut Harbour Expenses</u> (Basle: Prognos AG).

Roskam, A. K., C. J. Hoogenkiji, L. IJmker (1983), "Flow Metering in Dredging Systems," <u>Dredging and Port Construction</u>, February.

Roskam, A. K., C. J. Hoogendijk, L. IJmker (1983), "Flow Metering in Dredging Systems," <u>Dredging and Port Construction</u>, February.

Small Business Administration (1984), "Profile of the Dredging Industry for the Purpose of Setting a Size Standard," Washington, D.C.

Trawle, M. J. and J. A. Boyd, Jr. (n.d.), <u>Effects of Depth on Dredging Frequency, Report 1, Survey of District Offices</u> (Vicksburg, Miss: U.S. Army Engineer Waterways Experiment Station).

U.S. Army Corps of Engineers (1983), <u>Engineer Manual: Hydraulic Design of Deep-Draft Navigation Projects</u>, EM 1110-2-1613 (Washington, D.C.: U.S. Army Corps of Engineers).

U.S. Army Corps of Engineers (1965), <u>Engineer Manual: Tidal Hydraulics</u>, EM 1110-2-1607 (Washington, D.C.: U.S. Army Corps of Engineers).

Van Zutphen (1983), "Automation in the Dredging Industry," <u>Dredging and Port Construction</u>, February 1983.

Webster, W. C. (1983), "Steering and Maneuvering: State of the Art Report," Presentation to American Towing Tank Conference, Washington, D.C., August 1983.

9
Environmental Issues

Dredging and the disposal of dredged materials have the potential to cause physical and biological effects, and this potential, particularly when sediments removed by dredging are contaminated by toxic substances, has raised concerns about the environmental effects of dredging and disposal. Among the potential physical effects, implied in Chapter 8, is that a dredged channel or maneuvering area represents a change in the geometry of a tidal body of water, and local circulation and other patterns of flow are sensitive to such changes. Dredging and disposal activities directly disrupt bottom-dwelling communities; remove sediments from the bottom that may have collected toxic and other hazardous materials from upstream runoff and discharges; and transfer these sediments to other areas, with the possible consequence of mobilizing and dispersing the associated contaminants. These represent the potential physical and biological effects of greatest concern.

A great deal of research has been undertaken in the past decade to improve our understanding of the actual physical, biological, and public health implications of dredging and the disposal of dredged materials. This chapter reviews the accumulated knowledge and what it suggests for existing and future policies adopted to protect the marine and coastal environment, living marine resources, and public health.

SEDIMENTS

Deposits of sediments found within most ports can be divided into two primary classes: deep sediments, typically representing the major fraction forming the lower layers of the sediment column and known to have been in place for times that are long compared to the local history of industrialization; and surficial sediments, the more mobile fraction, found at or near the surface of the sediment column and typical of incoming sediments. The latter group includes the materials of primary concern for most dredging projects: The rate of deposition of surficial materials governs the extent and frequency of

117

maintenance dredging. The frequently elevated levels of contaminants in these sediments (as indicated by concentrations of oil and grease, trace elements, and long-lived synthetic organic compounds), leads to concerns about potential short- and long-term effects associated with mobilization, dispersal, and uptake of the contaminants from resuspension by dredging and from the disposal of dredged sediments. In contrast, deeper sediments are infrequently disturbed or displaced. They typically display a chemical composition that differs but slightly from the earth's average crustal materials in the drainage basin (Taylor, 1964), and exhibit little if any evidence of anthropogenic activities. These characteristics favor limited adverse effects following from the displacement of these materials.

Recent surficial sediments are composed of materials arriving from a number of sources by atmospheric and waterborne routes. From a mass-flux standpoint, waterborne inputs significantly exceed atmospheric contributions. The primary sources of waterborne materials include erosion of adjoining lands induced by rainfall and runoff, streambank and channelway erosion, biological activity in the water column and at the sediment-water interface, and (for coastal ports) the landward transport of sediments suspended over the adjoining continental shelf or the adjoining estuary. In addition to these natural sources, particulates are introduced to ports by a variety of anthropogenic activities, including the discharge of sewage effluent, industrial outfalls, direct dumping of debris, and discharges from street drainage and flood-control systems.

Although the relative importance of each of the sources supplying sediments varies from port to port, the significance of anthropogenic activity is clear. Agriculture and mining operations in rural areas and increased construction activity in urban areas in the past hundred years have raised the percentage of rainfall- and runoff-induced erosion products entering the waterways. Erosion from construction sites may yield 10 to 100 times the sediment produced per unit area by mining or agriculture (Gross and Palmer, 1979). Estimates suggest that within the northeastern U.S., construction activities to support an increase of 1000 in local population can result in the mobilization of approximately 600 to 1600 tons of sediment over an initial one- to five-year construction period (Wolman and Schick, 1967). The downstream transport of these materials, often from within immediate storage areas along the upper reaches of the waterway, can proceed over periods of decades and continue to affect the lower river and adjoining estuarine areas long after modifications in groundcover have reduced or eliminated sediment runoff from the construction site (Meade, 1980). Such persistent supplies, often subject to episodic erosion during storms, complicate efforts to reduce the frequency of maintenance dredging by controlling sediment inputs to harbor areas.

The spatial distribution of waterborne sediments is sensitive to a variety of transport factors, including the volume of freshwater discharge and the slope and channel cross-section within the riverine reach, the volume and characteristics of the tidal flow, the geometry of the estuarine basin, and the effects of winds and waves in the estuarine and coastal areas. Within the freshwater regions,

downstream transport dominates, and maximum flux of sediments is
associated with periods of peak streamflow. As a result, sediment
distributions in this region often display significant temporal
variability and relatively high degrees of sensitivity to the
placement and orientation of fixed structures. This sensitivity has
been used to reduce the downstream flux of sediment through the
construction of dams or similar sediment-retention structures, or to
flush materials from piers and mooring areas.

In the estuarine region, transport routes display both spatial and
temporal variability in response to varying streamflow and tidal
conditions. In most estuaries, mixing of fresh and salt waters
produces density distributions favoring net seaward movement of
near-surface waters and their suspended loads, and a corresponding net
landward displacement of near-bottom water and associated suspended
sediment. This circulation system favors retention within the estuary
of a large percentage of the solids that can settle (introduced either
upstream or within the adjacent offshore), with maximum deposition
occurring in the vicinity of the "null zone," or area in which the
near-bottom downstream movement of river water encounters the upstream
flow of seawater (Ippen, 1966). This convergence results in a
significant reduction in horizontal velocity and favors an increased
rate of deposition of suspended sediments. Changes in riverflow,
tidal flow, or cross-sectional geometry lead to a relocation of the
null zone. The positioning of port facilities relative to this null
zone represents an important determinant governing the frequency of
dredging required to maintain desired depths. Consideration of this
factor often provides at least partial explanation for the substantive
difference in the dredging frequency required to maintain one port as
compared to another despite both having apparently similar flow and
sediment supply characteristics.

The combination of factors affecting sediment transport within
coastal port facilities favors establishment of a controlling channel
depth representing a condition of equilibrium between flow-associated
transport energy and sediment supply. Dredging to increase water
depth beyond the controlling values disturbs this equilibrium by
modifying the flow regime and generally causes an acceleration in
deposition rates to force the system's return to equilibrium. With
the characteristic controlling depth for the majority of the U.S. port
facilities equalling 30 ft (10 m) or less, maintenance of the federal
navigational channels to depths approaching 45 ft (14 m) typically
requires dredging to establish the desired depth followed by a
continuing cycle of maintenance dredging to maintain channel depth and
to counter accelerating deposition as the system attempts to regain
equilibrium.*

---

*Significant in-channel forces may also be generated by the vessels
themselves, particularly larger vessels. The effects of all in-
channel forces for sedimentation can be estimated with a model and
local data (Hochstein, 1980).

Annually, dredging activities in the United States result in the removal of approximately 300 x 10^6 cubic meters of sediment. The largest-volume operations are in the southern states, where sediment yields are high because deep weathering produces a deep soil profile, and along the Mississippi River (Figure 7, Appendix G). The majority of these operations are classed as routine maintenance intended to remove deposits of surficial sediments. As a result, the displaced materials are dominantly clays and silts with lesser amounts of sand, and a moderate to high water content and organic fraction (Figure 8, Appendix G).

Approximately 20 to 25 percent of these materials are disposed of in ocean or ocean-fringing sites. The remainder is deposited within or adjacent to project areas or at less proximate inland sites. Dredged sediment dominates the materials dumped in the oceans of the United States (see table below). Along several areas of the continental shelf with large estuaries, the disposal of dredged materials represents the dominant mechanism for transporting sediments from the continent to the oceans (Goldberg, 1975; Gross and Palmer, 1979).

Ocean Dumping in the U.S. in 1983.

Waste Type	Amount (10^3 tons)
Dredged material	65,160
Industrial wastes	304.5
Sewage sludge	8,312
Construction debris[a]	0
Solid waste[a]	0
Explosives[a]	0
Wood incin.	31
Chemical incin.[a]	0
Total	73807.5

[a]While no materials in this category were dumped in 1983, they have been in prior years.
SOURCE: Dredged Material: U.S. Army Corps of Engineers. All other materials: U.S. EPA

The fine-grained nature of the majority of surficial sediments, in combination with their sedimentation history and associated exposure to the variety of anthropogenic inputs discussed can cause the chemical composition of this fraction to differ significantly from the deeper sediments and the more general average crustal materials (see Table 20, Appendix G). The variations in constituent concentrations, above those produced by natural inputs, can display significant variability in both quality and quantity, and can be expected to be highly site-specific.

In general, concerns about dredging and the disposal of dredged materials center on elevated concentrations of selected trace-elements, principally cadmium, mercury, and lead, and the synthetic organics, with recent emphasis on the polychlorinated biphenyls (PCB) and polynuclear aromatic hydrocarbons (PAH). Other constituents of concern include the nutrients, phosphate, nitrate, and ammonia, oil and grease, pathogenic microorganisms, and on occasion, the sediment itself. Because of the relatively large volumes of surficial sediments being dredged, the presence of elevated levels of these constituents prompted more stringent controls on dredging and disposal and the initiation of a variety of field and laboratory studies to assess the range of potential effects and to establish procedures to mitigate adverse effects.

DREDGING PROCEDURES

With the increasing incidence of sediment contamination by toxic compounds, a variety of advanced dredging systems has been developed. Mechanical systems employing closed buckets and hydraulic systems using skirted horizontal augers in shallow water and pneumatic pumps in deeper areas have been used, in combination with a variety of electronic, microprocessor-based, control and monitoring arrays, to dredge highly contaminated materials both in the U.S. and abroad. Studies have indicated that such systems have the potential to effect significant reductions in the turbidity associated with dredging while providing increased removal efficiency relative to the more conventional systems (Herbich and Brahme, 1983). Although such systems are finding general application in selected areas, notably Japan, their use is not widespread, and the majority of available dredges are "classic" or well-established systems. This situation appears to be the result of the conservative character of the dredging industry (Linssen and Oosterbaan, 1978); uncertainty about the future needs for advanced dredging techniques and the availability of the required funding; and the acceptability of conventional dredges for most projects. As detailed in Chapter 8, congressional action instructing the Corps of Engineers to increase the percentage of federal projects contracted to private firms stimulated the development of more modern, high-efficiency hopper dredges; similar improvements could be made in other dredging technologies if they were considered necessary.

DREDGED MATERIAL DISPOSAL PROCEDURES

Since the enactment of the variety of laws favoring reduction in the use of the ocean as a receiving area for wastes in the 1970s, the management philosophy governing disposal of dredged materials has emphasized selection of sites and procedures so as to minimize the dispersion of sediments discharged at offshore sites and to reduce the leakage of particulates and associated contaminants from alongshore containment sites. This containment policy was intended to (1) minimize the area in which adverse effects might occur; (2) complement evaluations of the adverse effects; and (3) permit possible future removal of the materials if the effects proved unacceptable. The selection of this protocol did not represent a universally held value judgment that in all cases containment was to be preferred to dispersal. The relative merits of containment versus dispersion remain a matter of continuing discussion (see, e.g., Rhoads, et al., 1978; Kamlet, 1981).

Satisfaction of the containment policy has been a continuing consideration in the selection and ultimate use of dredged material disposal areas. In the Great Lakes, this policy (as embodied in the River and Harbor Act of 1970), and consideration of the composition of the dredged materials and the chemical environment characterizing the open-water disposal areas, has resulted in the essential elimination of open-water disposal in favor of diked containment areas. Within the marine coastal region, diked structures are increasingly employed. Facilities are now in use for several ports, including Norfolk and Baltimore, and additional units have been proposed for Long Island Sound (U.S. Army Corps of Engineers, 1979).

In contrast to the care exercised in the design and specification of diked containment areas, procedures for their operation, and procedures for the selection and designation of ocean disposal sites appear haphazard. Prior to passage of the Ocean Dumping Act, approximately 160 sites were used for the disposal of dredged materials within the open coastal waters or inner continental shelf of the United States. The majority of these sites are on the Atlantic and Gulf coasts (see following table). Positioning and selection of

Regional Distribution of Disposal
Volumes and Sites

Total Volume Ocean Dumped ($10^6 m^3$)

	1976	1977	1978	1979
Atlantic	18	11	17	12
Gulf	24	10	15	36
Pacific	8	11	8	8
Total	50	32	40	56

Number of Active Dumpsites

	1976	1977	1978	1979
Atlantic	28	20	23	20
Gulf	20	18	23	16
Pacific	24	25	21	14
Total	72	63	67	50

From: Kamlet, 1983.

these sites was with few exceptions a simple function of proximity to the project area. Minimizing project costs favored locating disposal sites close to the dredging project. In 1977 (following enactment of the Ocean Dumping Act), the Environmental Protection Agency reduced the number of ocean sites from 160 to 127 (subsequently increased to 131), issued interim designations for each site, and initiated a series of investigations that was intended to lead to final designations (if appropriate) for the sites. With few exceptions, the sites retained their historical positions on the assumption that extending the effects of direct dumping to previously unused areas was unjustified in the absence of more detailed data.

The site-designation process for ocean disposal remains unfinished today, and the majority of the sites retain their interim designation. Owing to a series of legal settlements (Kamlet, 1983) and interagency agreements, the Environmental Protection Agency is committed to the timely completion of the designation process at 29 sites and has recently proposed a protocol to be used during these evaluations (Bierman and Reed, 1983). No final completion date has been established for the remaining sites.

Throughout the period of site designation, the disposal of dredged materials at open-water ocean disposal sites has continued. In the absence of site-specific data detailing dispersion and other important environmental characteristics, the operational criteria followed by the Corps of Engineers primarily emphasize the accurate placement of dredged material within the boundaries of the designated site. Procedures employing precision navigational systems (including loran-C) have been incorporated within routine disposal operations and detailed bathymetric surveys have been initiated at several sites to monitor the results. These survey data indicate that for the case of scow discharge of mechanically dredged materials, the consistent release of sediments at designated navigational coordinates or adjacent to a defined dumping buoy can produce coherent deposits of dredged material at specified points in the disposal area (see Figure 9, Appendix G). Similar results can be achieved with hydraulically dredged materials discharged from hopper dredges. Placement accuracy tends to degrade progressively for pipeline discharge of muds because of increasing water content or sediment fluidization and associated increased potential for dispersion. For coarser materials, however, even pipeline discharge can result in coherent placement of dredged materials.

The availability of precision navigation and high-resolution acoustic profiling systems permits the management of ocean disposal sites to a degree not previously attainable. In combination, these systems allow sequenced placement of dredged materials at a number of specified points within the disposal area, avoiding development of prominent mounds or shoals, and permit quantitative determination of the amounts of materials actually placed within the disposal site during a given project. These data allow estimates of the volume of material loss (but not necessarily of contaminant loss) occurring throughout the dredging and disposal operation and during the immediate post-disposal period as the materials settle and become

compacted. Such calculations assist both engineering and
environmental determinations, and in addition, provide a measure of
surveillance which serves to discourage the "short-dumping" or
"off-site" disposal practices that were common prior to 1970.
Finally, the development of precise placement procedures and
associated follow-up surveys promises to provide a means of reducing
the potential for biotic exposure or contaminant release from
contaminated dredged materials by allowing placement of a clean
"cover" or "cap" of sediments over these materials. This procedure is
discussed in a succeeding section ("The Disposal Area").

ENVIRONMENTAL EFFECTS

A large number of investigations have been carried out in the last 15
years to assess the environmental effects of dredging and
dredged-material disposal. These include (1) the Marine Ecosystems
Analysis (MESA) Program initiated in 1974 by the National Oceanic and
Atmospheric Administration (NOAA) with particular emphasis on the
disposal of wastes (including dredged materials) in the New York Bight
and lower New York Harbor (Ecological Stress..., 1982); (2) the
Dredged Material Research Program (DMRP) a five-year, $30-million
program mandated by Congress specifically to study the effects of
dredging and the disposal of dredged materials, and to develop
improved dredging systems and alternative disposal schemes (see U.S.
Army Corps of Engineers, 1980, for publications list); (3) a variety
of site-specific studies of dredging and the disposal of dredged
material often associated with the preparation of a required
Environmental Impact Statement (EIS), and studies by individual
divisions and districts of the Corps, such as the Disposal Area
Monitoring System (DAMOS) sponsored by the New England district to
permit continuing environmental evaluations of the active disposal
sites in the region (Science Applications Intl., 1984).

Reviews of the literature resulting from these investigations
provide reasonably clear indications of the short-term effects of
dredging and disposal activities, but often raise as many new
questions about long-term effects as they provide answers for old
ones. The data suggest that it is possible, using existing equipment
and procedures, to design and carry out a dredging project in which
the short-term effects are both minimal and acceptable. Specification
of the associated long-term effects is more difficult. This body of
information provides a useful first-order picture of the range of
environmental effects associated with dredging and disposal processes
and serves to highlight the areas needing further elaboration to
complement environmental management.

THE DREDGING SITE

Of the large number of studies intended to detail the environmental
effects of dredging and disposal, a relatively small percentage have

focused on the dredging site itself. The studies that have been conducted in this area have placed primary emphasis on the extent and character of the sediment resuspension induced by dredging and the influence of these materials on local pelagic fish populations, or the benthic community found in the areas adjacent to the channel being dredged, or both. Additional studies have examined the effects of dredging-induced resuspension on local water quality, with particular emphasis on the release of particulate-associated contaminants, and have detailed the extent to which dredging affects local circulation and sediment transport by modifying channel depth and cross-sectional characteristics. Data from these studies provide a basis for the development of quantitative predictive models.

Both mechanical and hydraulic dredging operations introduce significant quantities of sediment into the water column immediately adjacent to the operating dredge. For mechanical operations in areas of moderately fine-grained cohesive sediments, concentrations of suspended materials adjacent to the dredge have been observed to exceed background levels by more than two orders of magnitude, as shown in Figure 10 (Appendix G). Similar variations have been observed adjacent to an operating cutterhead dredge with concentrations varying as a function of the size and relative production of the dredge (Figure 11, Appendix G). Hopper dredge overflows appear to have the potential to produce the maximum perturbation of suspended material: observations at several locations indicate concentrations adjacent to the overflow port in excess of 100 gm/l, or more than five orders of magnitude above background (Figure 12, in Appendix G).

The materials suspended by the operating dredge are distributed downstream by the local transport field, and display concentrations varying as a function of mass-settling properties, free-stream velocity, and associated turbulent diffusion characteristics. Observations indicate that for representative estuarine conditions, this combination of factors favors rapid deposition of the resuspended materials. The sediment plume represents a relative near-field feature displaying characteristic longstream spatial scales of less than 2000 m (see Figures 13 and 14, in Appendix G, for example). Comparisons between distributions observed at a variety of sites, and for several different dredge systems, indicate clear similarity and have permitted development of reasonably accurate predictive modeling requiring only definition of the initial concentrations adjacent to the dredge and an estimate of free-stream diffusion and particulate settling velocities (Cundy and Bohlen, 1980). These models have proved useful for evaluation of the potential effects of dredging.

In addition to the solid particulate phase, the operating dredge also directly and indirectly alters the concentrations of dissolved nutrients and selected trace elements within the waters in the immediate vicinity of the dredge. Studies of these constituents indicate elevated concentrations above background within an area representing approximately 30 percent of the total suspended material plume. Over the remaining area of the plume, dilution and particle scavenging favor a return to background levels (Tramontano and Bohlen, 1984).

The limited spatial extent of the suspended material plume produced by the typical estuarine dredging operation effectively limits the associated effects to areas immediately adjacent to the operating dredge. Within this region, the elevated suspended material concentrations serve to (1) increase turbidity, which reduces the penetration of light and associated photosynthetic activity; and (2) provide a continuing supply of sediment for deposition along and over adjoining benthic areas. The potential effects associated with these material concentrations appear to be limited by a combination of factors. Within the water column, the effects of particulates on the drifting biotic community, including zooplankton--although difficult to evaluate--are considered negligible because of the limited area affected and the characteristically short exposure time. For the more mobile, free-swimming organisms, potential effects are further reduced by their ability to avoid the resuspension area. The benthic biological community not affected directly by dredging can be affected by the rain of resuspended sediments. The rapid settling of these materials serves to confine the primary effects to the immediate vicinity of the operating dredge, resulting in zones of influence having characteristic spatial scales ranging from 100 to 1000 m^2. The deposition of suspended sediments within this area affects particularly the filter-feeding organisms, including several species of commercial value such as oysters, scallops, and blue mussels. The extent and character of the effects varies as a function of the concentration levels of suspended sediments, sedimentation rate, and exposed species. Persistent concentrations in excess of 2 gm/l, or deposition sufficient to produce deep burial (>20 cm), or both, can prove lethal to a majority of benthic organisms. Such conditions, however, exist only within the areas immediately adjacent to the operating dredge where the effects are generally negligible compared to those induced directly by the bucket or hydraulic intake. Beyond this area, over the undisturbed region flanking the dredged channel, the increase induced by dredging in suspended material concentration over background seldom exceeds 100 mg/l, typically representing a potential deposition of less than 1 cm. In these conditions, the ultimate effects become primarily a function of the tolerance of the exposed species. Epifaunal suspension feeders such as oysters and mussels display maximum sensitivity. A variety of investigations has shown that these organisms as adults can tolerate suspended material concentrations in the range of 100 to 1000 mg/l over reasonably short exposure times and that on occasion such exposure can serve to stimulate pumping activity and increase growth by increasing nutrient supplies (Lunz, 1938; Loosanoff and Tomars, 1948; Loosanoff, 1961; Stern and Stickle, 1978). Nevertheless, persistent exposure to high concentrations of suspended sediments, or shallow burial (<1 cm), or both, is generally lethal (Kranz, 1974). For the larval and juvenile stages of these organisms, effects appear to be negligible at concentrations below 200 mg/l, and slowly increase to critical at approximately 750 mg/l (Davis and Kidu, 1969). Although such concentrations occur only in the immediate vicinity of the dredge, the degree of uncertainty in the available data on the biological effects of those concentrations appears sufficient to justify the management practices applied in many areas limiting dredging activity during the critical spawn-and-set periods of the commercially valuable species of shellfish. Restrictions based on finfish sensitivity, however, appear to be seldom justified, except perhaps if the channel and dredge occupy a large fraction of the waterway's cross-section, and the waterway is a major passage for migrating species.

An additional factor often limiting the environmental effects of dredging is the natural degree of variability in the sediment transport system of the majority of shallow-water lakes, estuaries, and coastal embayments, as well as inland waterways. In many estuaries, near-bottom concentrations of suspended material vary by more than an order of magnitude over each six-hour half-tidal cycle (where there are semi-diurnal tides) as fine-grained organic and inorganic materials are alternately suspended or deposited in response to the varying tidal velocities (Meade, 1972). Over longer periods, the suspended material field within each of these systems will be perturbed aperiodically by short-term, high-energy events sufficient to increase concentrations by several orders of magnitude above background. Such events display a typical recurrence interval of less than twelve months and often represent the primary determinant governing the flux of sediments to a given system and through it. Less-frequent events can have major effects on coastal sedimentary systems. The effects of tropical storm Agnes on the sedimentary system of Chesapeake Bay present a particularly clear illustration of the potential of these less-frequent, aperiodic events (Schubel, 1974; Zabawa and Schubel, 1974). Perturbations occurring over a range of temporal scales will each tend to affect significantly larger areas than those affected by routine dredging operations. This factor, in combination with the amount of sediment displaced by events suggests that against such perturbations, the system-wide influence of sediment suspension produced by dredging will generally be negligible (Bohlen, 1980).

In addition to the variety of relatively short-term effects, dredging operations may induce a number of longer-term effects associated primarily with modifications in local circulation and sediment transport following changes in channel depth and cross-sectional area. These effects are most likely to be significant within estuarine areas, where altered channel contours can increase the degree of salinity intrusion and alter vertical mixing, leading to a modification in the density structure and associated gravitational circulation, and causing repositioning of the areas of maximum sediment accumulation (Simmons and Brown, 1969). Changes in mixing and gravity circulation can also affect the distribution of dissolved oxygen and other water-quality parameters.

The relationships between changes in channel geometry and changes in circulation and channel shoaling have been detailed in a variety of investigations (Harleman and Ippen, 1969). The investigations indicate that while modification in channel configuration has the potential to alter local circulation characteristics, the physical effects can be predicted with reasonable accuracy using appropriate hydraulic and numerical models (Thatcher and Harleman, 1972; Festa and Hansen, 1976; Officer, 1980).

Increasing salinity intrusion by channel deepening may lead to encroachment of salt into local supplies of groundwater and surface water. Increases in salinity associated with channel deepening may affect the viability of adjoining freshwater wetlands and tidal marshes, which may in turn influence local aquatic resources, including the range of freshwater, anadromous, estuarine, and coastal fish populations. These implications appear to be of particular concern in the Gulf of Mexico, where they have been the subject of discussion for more than 30 years (Morgan, 1973). There is no doubt that undesirable effects can accompany salinity intrusion in specific conditions. These are associated primarily with new construction dredging, and not with routine maintenance dredging. As a result, plans proposing major alterations in channel depth and cross-section should include consideration of the associated modification in salinity intrusion in sufficient detail to permit resolution of changes induced by dredging and the short-term natural variations associated with fluctuations in river flow and astronomical and meteorological tides (Morgan 1973).

The character and extent of biological recolonization within the dredged channel varies as a function of the post-project hydrographic and sedimentological conditions, and the frequency of dredging. A significant increase in salinity above preproject levels and an associated increase in sedimentation rates, particularly of the finer-grained materials, will favor a permanent modification in the composition of the benthic community, a possible shift to more salinity-tolerant organisms, reductions in diversity, and slow rates of initial recolonization (Kaplan et al., 1974; Taylor and Saloman, 1968). The rates at which these alterations proceed vary substantially from region to region: times for reestablishment of a stable community range from 1.5 to 12 years. In some areas, recovery times are long compared to dredging intervals, resulting in a continuing state of instability within the benthic community. The over-all result of these variations for estuarine productivity has not been demonstrated and appears negligible in most cases, owing to the small areas affected. For large new construction dredging projects that would significantly alter channel cross-sectional areas, the potential for such changes should be carefully assessed.

The Disposal Area

Upland Sites and Sites Fringing the Shoreline

Since the initiation of dredging in the United States in the late 1800s, upland sites and sites fringing the shoreline have been primary receiving areas for dredged materials. Materials placed in these areas have served as construction fill for airports, footing for recreational areas and flood-control structures or dikes; and for the coarser fraction, as replenishment sands for beachfront restoration.

In recent years the use of such areas for disposal has tended to decrease because of increasing population pressure and the resultant decrease in available open space, legislation prohibiting the filling of wetlands and marshes, and concerns about the release of contaminants associated with dredged materials. As a result of these factors, a reasonably coherent policy concerning the use of terrestrial sites has evolved sufficient to effect a general elimination of the haphazard disposal of dredged materials that had been common in many areas. As implemented, this policy favors the use of fringing or upland sites if secondary benefits can be realized--the construction of a tidal marsh, creation of a wildlife habitat, or beach replenishment--or if the degree of contamination exceeds established levels for open-water disposal. Marsh and habitat development with uncontaminated sediments and beach replenishment have been studied in some detail and shown to have relatively short-lived adverse effects, these occurring principally during the placement operation (Lunz et al., 1978). Determining the adverse effects associated with terrestrial disposal of contaminated materials and the advisability of using land sites rather than open-water sites is more difficult and controversial.

Arguments favoring the use of terrestrial sites as receiving areas for contaminated dredged materials emphasize the combination of containment, the ability to observe closely any negative effects, and the relative ease with which corrective actions, such as removal and relocation, could be taken if unacceptable effects are observed. Countering arguments point to the inherent difficulty of realizing absolute containment of dredged materials and the enhanced potential for release and mobilization of a variety of contaminants associated with placement of anaerobic sediments in an aerobic environment. The increased availability of oxygen results in the alteration of the phase of some sediment-associated heavy metals from the insoluble sulfide form (favored in reducing conditions) to more soluble sulfates (Kester et al., 1983). In addition, these reactions affect the pH of the interstitial waters generally leading to more acidic conditions and the potential for additional release of particulate-bound contaminants. The extent and character of contaminant release resulting from this combination of oxidation reactions varies as a function of the redox potential (Eh) and pH. Increasing Eh and an associated decrease in pH relative to natural in situ values appears to favor release of a progressively wider range of trace metals (Gambrell et al., 1976). The potential for contaminant release from dredged materials placed in terrestrial sites and the associated probability of surface water or groundwater contamination, as well as increased availability to the local biological community complicates the management of these sites both during and after receipt of contaminated sediments.

Effective leachate control presumably can be achieved by the placement of impermeable liners to contain the materials and the use of settling and retention basins sufficient to permit evaporation or effective depuration. These procedures are expensive and significantly increase the area required for a containment site. The often equivocal nature of the effects that can be directly associated with all but the most toxic materials complicates justification of these added costs. Moreover, success in achieving and maintaining total leachate control has been marginal.

If a terrestrial containment site is used, it must be chosen carefully and should not be located in an unsuitable area such as atop an aquifer, in a wetland, or in an area of high runoff (Gordon et al., 1982). To the extent possible, the soil beneath the site should be predominantly fine-grained material to ensure a chemical capacity to adsorb and bind contaminants to particles, and be of high porosity and low permeability.

The best strategy for the disposal of contaminated dredged material is one that contains the particles, confines the contaminants to the particles and isolates the deposit and associated contaminants from plants and animals, and particularly from man. These conditions can perhaps be approached most closely by burial beneath the seafloor (Bokuniewicz, 1983), under a cap of clean sediment (Morton, 1983). All the major elements of a subaqueous burial operation have been demonstrated in the field including the intentional construction of a compact deposit (e.g., Morton, 1983; Bokuniewicz 1982) and the successful capping of fine-grained dredged sediment under a sand cap (e.g., Morton 1983; O'Connor, 1982). Indeed, a small operation to bury contaminated dredged mud in a submarine pit under a sand cap has been successfully completed (Sumeri, 1984). Available field studies and continuing laboratory tests indicate that the caps are apparently effective in containing contaminants (O'Connor, 1982; Brannon et al., 1984). Although the limiting criteria for a successful burial operation are not well known, a successful large-scale operation could be carried out so long as the conditions, materials, and techniques are not significantly different from those of the capping operations that have already been completed. Before the burial options could be routinely used in a wide range of conditions and materials, however, generally applicable criteria need to be developed concerning, for example, the spread of dredged sediments along the seafloor during the discharge process, the geotechnical conditions that allow capping, and the migration mechanisms of specific contaminants.

It is probably neither possible nor appropriate at this time to conclude categorically that either upland containment or subaqueous disposal is universally preferable for the management of contaminated dredged materials. As was pointed out by a Corps of Engineers scientist (Engler, 1981) following the DMRP, "containment of highly contaminated or toxic dredged material (at an upland disposal site) ...can be an environmentally sound and preferred alternative,but cannot be categorically considered better than (other management or disposal techniques)...."

The best, most appropriate, choice of a subaerial or a subaqueous disposal site will vary with the quantity and quality of material to be disposed of, the characteristics of the terrestrial and aquatic environments in that region, the uses society makes of these environments, and the availability of sites.

Open Water Sites

The placement of dredged materials in open-water disposal sites has the potential to induce a variety of short-term, acute, and longer-term, chronic environmental effects. The short-term effects are confined to the period of disposal and result primarily from direct burial of marine organisms or their exposure to increased concentrations of suspended materials, trace elements and other contaminants, and nutrients. The majority of these effects can be reduced or eliminated by proper site selection and project timing. Studies of longer-term effects have considered rates of recolonization and the character of the subsequent biological community, variations in contaminant body burdens within these organisms, reproductive success, and a variety of sublethal but persistent effects, such as alterations in genetic structure. This latter set of effects is by far the most difficult to assess, and consequently, is the least well known.

As in the case of dredging-induced resuspension, a number of field studies have shown that the open-water disposal of dredged materials by hydraulic pipeline or hopper barge produces increases in suspended-material concentrations that are short-lived, and that the primary effects of these short--lived increases are confined to the immediate vicinity of the discharge point. During hydraulic placement of materials by an outfall pipe, suspended-material concentrations vary as a function of mean grain size and production rate, with values decreasing rapidly with distance downstream. Typically, the perturbed suspended-material concentrations return to background within approximately 2000 m of the point of discharge (Figures 10 and 12 in Appendix G), and within a few hours after the discharge operation ends.

The discharge of materials from a hopper or scow creates a descending jet of sediment with a trailing wake of entrained waters and suspended particulates (Figure 15 in Appendix G). The water-column distributions of these latter materials will vary as a function of the sediment mass characteristics, particularly the degree of cohesion, and for water depths in excess of 100 m or so, the density structure of the water column. On impact with the bottom, a fraction of the descending mass will be redirected upwards, and an additional volume of sediment will be introduced into suspension from disturbance of the bottom. The energies associated with the combination of descending and ascending sediments then slowly dissipate and the cloud of materials settles toward the sediment-water interface. In water depths of approximately 20 to 50 m, this process typically results in a well-defined pile of sediment having a conical core and displaying symmetrical axial dimensions equal to approximately 30 percent of the water depth (Gordon, 1974). Investigations have shown that the distributions of suspended sediments resulting from both hydraulic discharge and barged disposal can be predicted reasonably well by analytical models (Koh and Chang, 1974; Wilson, 1979).

The sediments suspended during disposal operations have the potential to produce the same range of effects as sediments resuspended by the operating dredge. Although the potential is greater, the majority of the effects produced by ocean disposal of dredged material are considered negligible, except in areas dominated by sensitive species such as corals, or filter-feeding organisms such as oysters, clams, and mussels. Efforts are generally made in the selection of disposal sites to avoid sensitive areas, including those that support submerged aquatic vegetation and significant concentrations of commercially important shellfish.

The direct burial of the variety of benthic organisms resident within the disposal area represents the primary short-term environmental effect of dredged material disposal in open water. With few exceptions, organisms buried during large-volume disposal operations will not survive, resulting in nearly azoic conditions on completion of the project. Colonization of the dredged-material pile begins within a relatively short time, producing initially a benthic community displaying limited diversity and dominated by opportunistic, stress-tolerant species (Rhoads et al., 1978). Times associated with the development of this assemblage are typically short, ranging from weeks to less than a year. The rate and degree of subsequent change varies with the nature of the sediment, particularly its texture and cohesiveness; the relief of the mound above the seafloor and the sediment transport field. This combination of factors results in significant variability in substrate characteristics and benthic communities. Times associated with establishment of an equilibrium community vary from months to years (Obrebski and Whitlatch, 1981).

Beyond the obvious mortality produced by initial burial, the adverse environmental effects of dredged material disposal cannot be specified. The presence of the dredged material can alter local fish habitat, resulting in a local shift of the dominant species. Available data suggest, however, that while deposits of dredged material may inconvenience local fisherman, they do not necessarily reduce total yield or the landed value of commercial species (Chesapeake Biological Laboratory, 1970; Oppenheimer, 1984). Mounds of dredged material can, for example, interfere with nets that are towed or set to drift at specific depths. Some investigations suggest that the disturbance of the equilibrium state produced by some amount of dredged material disposal increases productivity, and can on the whole, be beneficial. These results form the basis for a recent proposal to test modification of the prevailing scheme (based on a small number of relatively large-volume dumpsites) to one favoring a larger number of smaller sites distributed throughout the estuary, or offshore, or both (Rhoads et al., 1978). The similarity between the proposed scheme and the spatial distribution of disposal areas prevailing prior to 1970, although obviously sited for substantially different reasons, raises some interesting questions concerning the optimum management of dredged-material disposal in estuaries and open coastal waters.

Coincident with the physical and biological variations occurring during and immediately after the disposal operation are a number of chemical processes that affect the distribution and ultimate bioavailability of the variety of organic and inorganic compounds associated with the dredged materials. Since many of these materials are known to be potentially toxic, the character and extent of chemical processing typically receives particular attention in efforts to detail the effects of disposed materials. A number of studies, representing a major portion of the effort to determine the environmental effects of dredged material disposal have considered contaminants found within both dissolved and particulate phases. The general approach used in both laboratory and field studies has been to establish a reference or control (station or sample), if possible, and to collect some series of pre-project baseline data, and then with the onset of disposal, to initiate analyses comparing disposal-site conditions to those prevailing in the control.

Reviews of the large body of literature resulting from these investigations indicate general agreement that the availability and ultimate biological uptake is higher for contaminants associated with the dissolved phase than for those found within the particulate phase. This availability is associated primarily with the release of interstitial waters, and favors maximum uptake during and for some short time after the completion of the disposal operation. The subsequent effects vary with a variety of factors, including time of year, class and age of the organism, and the particular contaminant(s). The principal adverse effects are generally associated with well-known contaminants, including halogenated hydrocarbons, such as PCB, and mercury (see Table 21 in Appendix G).

Beyond this class of essentially short-term effects associated with dissolved-phase uptake, evaluations rapidly become more difficult. Considerations of particulate-phase contaminants often show weak correlation between sediment concentrations and body burden levels within the local biological community (Pequegnat, 1983). A variety of studies conducted during the DMRP both in the laboratory and the field provided similar results and lead to the conclusion that for short-term effects "...impacts of dredged materials are primarily associated with physical effects and....biochemical interactions are infrequent and bioaccumulation of metals and hydrocarbons negligible" (Engler, 1981). The data imply that the availability of the contaminants associated with the particulate phase is limited by electrochemical binding that requires major changes in pH or Eh for dissociation (Gambrell et al., 1976). For all but the most severe contamination involving moderately to highly toxic materials, short-term biological effects are essentially limited.

Despite the large body of data developed by the DMRP supporting these conclusions, acceptance of the minimal-effect view is far from widespread. Our conclusion based on review of these data, as well as the variety of information developed within other programs (MESA, DAMOS, etc.), is consistent with the view, but additional, more sophisticated, and longer-term studies are required for unequivocal assessment. Until such information is available, an environmentally conservative course appears prudent. The determination of uptake of contaminants and ultimate biological effects are both complicated by a variety of fundamental unknowns--the factors governing an adequate control or reference station; the life histories of the selected indicator organisms; the mechanisms used by the indicator organism to metabolize contaminants; and the physiological effects of continued exposure to toxic contaminants, including consideration of genetic modifications. Compounding the difficulties associated with these unknowns is the high degree of variability associated with the inshore biological community (Livingston, 1982). This combination of factors generally precludes simple determination of cause and effect using short-term data sets. Based on these factors, the prevailing opinion among experts is that the effects associated with long-term exposure to moderate or low levels of contamination are, for the majority of the marine biological community, largely unknown and that therefore any potential for adverse effects should be minimized through proper management practices based on the best available information.

Regulatory Procedures

Environmental legislation and regulation are discussed in Chapter 7. From the environmental standpoint, the primary difficulties associated with procedural and institutional matters are the lack of responsiveness to the flow of information about environmental effects--both positive and negative--and lack of assessment of the implications for present criteria. In the case of dredging and dredged material disposal, it appears that far more is known about environmental effects and probable causes than is incorporated in regulatory criteria and environmental practices. Streamlining the regulatory process has the potential to improve not only port management but also the incorporation of scientific results in environmental criteria.

SUMMARY

Port dredging and disposal operations have the potential to induce a variety of short- and long-term environmental effects. The majority of these effects can be predicted, and efforts are proceeding to resolve the remaining unknowns. Even within the category of unknown effects, sufficient data exist to permit definition of the potential range of effects that might occur in extreme conditions and to select management strategies that minimize the probability of adverse effects. Overall, the effects associated with a proposed dredging project can be reasonably well defined and controlled. This review suggests that the major concerns remain with the disposal of contaminated sediments containing moderate to high concentrations of toxic materials. Since typically this contaminated fraction constitutes a relatively small percentage of the materials removed during maintenance of existing berths, channels, and maneuvering areas, and an even smaller percentage of the sediments associated with new construction dredging, their presence should not represent a major impediment to future port management or development plans if dredging and disposal methods can be matched to their location, type, and amount.

REFERENCES

Barnard, W. D. (1978), "Prediction and Control of Dredged Material Dispersion Around Dredging and Open-Water Pipeline Disposal Operations," Tech. Report D5-78-13, Dredged Material Research Program, U.S. Army Engineer Waterways Experiment Station, Vicksburg, Miss.

Bierman, V. J., Jr. and M. Reed (1983), "Proceedings of a Workshop for the Development of a Scientific Protocol for Ocean Dumpsite Designation," Applied Science Associates, Wakefield, Rhode Island.

Bohlen, W. F. (1980), "A Comparison Between Dredge Induced Sediment Resuspension and that Induced by Natural Storm Events," Proc. 17th Coastal Eng. Conf. (New York: American Society of Civil Engineers), pp. 1700-1707.

Bohlen, W. F., D. F. Cundy, and J. M. Tramontano (1979), "Suspended Material Distributions in the Wake of Estuarine Channel Dredging Operations," Estuarine and Coastal Mar. Sci., 9: 699-711.

Bokuniewicz, H. J. (1983), "Submarine Borrow Pits as Containment Sites for Dredged Sediment," Wastes in the Ocean, Vol. 2, Dredged Material Disposal in the Ocean, D. R. Kester, et al., eds. (New York: Wiley & Sons) pp. 215-227.

Bokuniewicz, H. J. (1982), "Burial of Dredged Sediment Beneath the Floor of New York Harbor, Oceans '83 (Washington, D.C.: Marine Technology Society), pp. 1016-1020.

Brannon, J. M., R. E. Hoeppel and D. Gunnison (1984), "Efficiency of Capping Contaminated Dredged Material," Dredging and Dredged Material Disposal, Vol. 2, R. L. Montgomery and J. W. Leach, eds. (New York: American Society of Civil Engineers), pp. 664-673.

Chesapeake Biological Laboratory (1970), "Gross Physical and Biological Effects of Overboard Spoils Disposal in the Upper Chesapeake Bay," Special Report #3, Contribution #397, Natural Resources Institute, University of Maryland, College Park, Md.

Council on Environmental Quality (1970), Ocean Dumping: A National Policy (Washington, D. C.: Government Printing Office).

Cundy, D. F. and W. F. Bohlen (1980), "A Numerical Simulation of the Dispersion of Sediments Suspended by Estuarine Dredging Apparatus," Estuarine and Wetland Processes, P. Hamilton and K. B. MacDonald, eds. (New York: Plenum Press).

Davis H. C. and H. Kidu (1969), "Effects of Turbidity Producing Substances in Sea Water on Eggs and Larvae of Three General of Bivalve Mollusks," The Veliger, 11: 316-323.

Ecological Stress in the New York Bight: Science and Management (1982), Gary F. Mayer, ed. (Columbia, S.C.: Estuarine Research Federation).

Engler, R. M. (1981), "Impacts Associated With the Discharge of Dredged Material: Management Approaches," Use of the Ocean for Man's Wastes: Engineering and Scientific Aspects, (Washington, D. C.: National Academy Press) pp. 129-185.

136

Environmental Protection Agency and U. S. Army Corps of Engineers
(1977), "Ecological Evaluation of Proposed Discharge of Dredged
Material Into Ocean Waters," EPA/COE Technical Committee on
Criteria for Dredged and Fill Material, U.S. Army Engineer
Waterways Experiment Station, Vicksburg, Miss.

Festa, J. F. and D. V. Hansen (1976), "Effects of Dredging and River
Flow Diversion on Estuarine Circulation," Estuarine and Coastal
Mar. Sci., 4: 309-324.

Gambrell, R. P., R. A. Khalid, and W. H. Patrick, Jr. (1976),
"Physiochemical Parameters That Regulate Mobilization and
Immobilization of Toxic Heavy Metals," Dredging and Its
Environmental Effects, P. A. Krenkel, J. Harrison, and J. C.
Burdick III, eds. (New York: American Society of Civil Engineers),
pp. 418-434.

Goldberg, E. D. (1975), "Man's Role in the Major Sedimentary Cycle,"
The Changing Global Environment, S. Fred Singer, ed. (Dordrecht,
Holland: Reidel Publishing Co.), pp. 275-294.

Gordon, R. B. (1974), "Dispersion of Dredge Spoil Dumped in Near
Shore Waters," Estuarine and Coastal Mar. Sci., 2: 349-358.

Gross, M. G. and H. D. Palmer (1979), "Waste Disposal and Dredging
Activities: The Geological Perspective," Ocean Dumping and Marine
Pollution, H. D. Palmer and M. G. Gross, eds. (Stroudsbeck, Penn.:
Dowden, Hutchinson and Ross, Inc.), pp. 1-7.

Harleman, D. R. F. and A. T. Ippen (1969), "Salinity Intrusion
Effects in Estuary Shoaling," Proc. ASCE, Hyl, 95: 9-27.

Herbich, J. B. and S. B. Brahme (1983), "Literature Review and
Technical Evaluation of Sediment Resuspension During Dredging,"
Report No. COE-266, Texas A&M Research Foundation, College Station,
Texas.

Hochstein, A. (1980), "Analysis of the Effect of Tow Traffic on the
Physical Components of the Environment," Report prepared for the
Department of the Army, Huntington District, Corps of Engineers, by
Louis Berger and Associates, Inc. In U.S. Army Corps of Engineers,
Huntington (W. Va.) District (1981), Gallipolis Locks and Dam
Replacement, Ohio River, Phase 1, Advanced Engineering and Design,
General Design Memorandum, Appendix J: "Environmental and Social
Impact Analysis, Part 1.

Ippen, A. T., ed. (1966), Estuary and Coastline Hydrodynamics (New
York: McGraw Hill).

Kamlet, K. S. (1981), "The Oceans As Waste Space: The Rebuttal,"
Oceanus, 24: 10-17.

Kamlet, K. S. (1983), "Dredged-Material Ocean Dumping: Perspectives
on Legal and Environmental Impacts," Wastes in the Ocean, Vol. 2,
Dredged-Material Disposal in the Ocean, D. R. Kester et al., eds.
(New York: Wiley & Sons), pp. 29-70.

Kaplan, E., J. Welker, and M. Draus (1974), "Some Effects of Dredging
on Populations of Macrobenthic Organisms," Fish. Bull., 72: 445.

Kester, D. R., B. H. Ketchum, I. W. Duedall, and P. K. Park (1983),
"Have the Questions Concerning Dredged Material Disposal Been
Answered?" Wastes in the Ocean, Vol. 2, Dredged Material Disposal
in the Ocean, D. R. Kester et al., eds. (New York: Wiley & Sons),
pp. 276-287.

Koh, R. C. Y. and Y. C. Chang (1974), "Mathematical Model for Barged Ocean Wastes," Report prepared for Environmental Protection Agency, National Environmental Research Center, Corvallis, Ore.

Kranz, P. M. (1974), "The Anastrophic Burial of Bivalves and its Paleo-ecological Significance," J. Geology, 82: 237-265.

Linssen, J. G. Th. and W. Oosterbaan (1978), "Dredging Equipment: Its Past Performance and Future Development," Terra et Aqua, 16: 1-13.

Livingston, R. J. (1982), "Long-Term Variability in Coastal Systems: Background Noise and Environmental Stress," Ecological Stress and the New York Bight: Science and Management, G. F. Mayer, ed. (Columbia, S.C.: Estuarine Research Federation), pp. 605-620.

Loosanoff, V. L. (1961), "Effects of Turbidity on Some Larval and Adult Bivalves," Gulf and Caribbean Fisheries Inst. Proc., 14: 80-95.

Loosanoff, V. L. and F. D. Tomars (1948), "Effect of Suspended Silt and Other Substances on Rate of Feeding of Oysters," Science, 107: 69-70.

Lunz, J. D., R. J. Diaz, and R. A. Cole (1978), "Upland and Wetland Habitat Development with Dredged Materials: Ecological Considerations," Tech. Report DS-78-15, DMRP, U.S. Army Engineer Waterways Experiment Station, Vicksburg, Miss.

Lunz, R. G. (1938), "Part 1. Oyster Culture With Reference to Dredging Operations in South Carolina," Report to the U.S. Engineer Office, Charleston, S.C.

Meade, R. H. (1980), "Man's Influence on the Discharge of Fresh Water Dissolved Material and Sediment by Rivers to the Atlantic Coastal Zone of the United States," River Inputs to Ocean Systems, Proc. SCOR Workshop, 26-30 June 1979, Rome (Paris: UNESCO), pp. 13-17.

Meade, R. H. (1972), "Transport and Deposition of Sediments in Estuaries," Geological Soc. of America Mem., 133: 91-120.

Morgan, J. P. (1973), "Impact of Subsidence and Erosion on Louisiana Coastal Marshes and Estuaries," Proc. Coastal Marsh and Estuary Management Sym., Louisiana State University, Baton Rouge, Louisiana, pp. 217-233.

Morton, R. W. (1983), "Precision Bathymetric Study of Dredged-Material Capping Experiment in Long Island Sound," Wastes in the Ocean, Vol. 2, Dredged Material Disposal in the Ocean, D. R. Kester et al., eds. (New York: Wiley & Sons), pp. 99-124.

Obrebski, S. and R. B. Whitlatch (1981), "Effects of Dredging on Benthic Communities in the San Francisco Southwest Outfall Areas," Report prepared for the San Francisco Clean Water Program, Tomales Bay Marine Laboratory, Marshall, California.

O'Connor, J. M. and J. W. Rachlin (1982), "Perspectives on Metals in New York Bight Organisms: Factors Controlling Accumulation and Body Burdens," Ecological Stress and the New York Bight: Science and Management, G. F. Mayer, ed. (Columbia, S.C.: Estuarine Research Federation), pp. 655-673.

O'Connor, J. (1982), "Evaluation of Capping Operations at the Experimental Mud Dump Site, N.Y. Bight Area, 1980," Final Report #DACW 39-82-2544, U.S. Army Corps of Engineers, New York District.

Officer, C. B. (1980), "Box Models Revisited," Estuarine and Wetland Processes, P. Hamilton and K. B. Mac Donald, eds. (New York: Plenum Press), pp. 65-114.

Oppenheimer, C. H. (1984), "Environmental Effects of Dredged Material," Commissioned paper for Committee on National Dredging Issues, February 1984.

Pequegnat, W. E. (1983), "Application of Classification Criteria to Dredged Material With Emphasis Upon Petroleum Hydrocarbons and With Additional Consideration of Lead in Dredged Material," Report prepared for International Association of Ports and Harbors.

Pequegnat, W. E., D. D. Smith, R. M. Darnell, B. J. Presley, and Robert O. Reid (1978)," An Assessment of the Potential Impact of Dredged Material Disposal in the Open Ocean," Tech. Report D-78-2, DMRP, U.S. Army Engineer Waterways Experiment Station, Vicksburg, Miss.

Rahn, K. A. (1976), "The Chemical Composition of the Atmospheric Aerosol," Technical Report, Graduate School of Oceanography, University of Rhode Island, Kingston, R. I.

Rhoads, D. C., P. L. McCall, and J. Y. Yingst (1978), "Disturbance and Production on the Estuarine Seafloor," Am. Scientist, 66: 577-586.

Schubel, J. R., D. J. Hirschberg, D. W. Pritchard, and M. G. Gross (1980), "A General Assessment of Selected Dredging/Disposal Options for Three Federal Dredging Projects in Upper Chesapeake Bay," Special Report No. 40, Marine Sciences Research Center, State University of New York at Stony Brook.

Schubel, J. R. (1974), "Effect of Tropical Storm Agnes on the Suspended Solids of Northern Chesapeake Bay," Suspended Solids in Water, R. J. Gibbs, ed. (New York: Plenum Press), pp. 113-132.

Science Applications Intl. Corp. (1984), "Disposal Area Monitoring System, Summary of Program Results 1981-1984," R. W. Morton, J. H. Parker, and W. H. Richmond, eds., Contribution #46 to Northeast Division, U.S. Army Corps of Engineers, Waltham, Mass.

Simmons, H. B. and F. R. Brown (1969), "Salinity Effects on Estuarine Hydraulics and Sedimentation," Int. Assoc. for Hydr. Res., Proc. 13th Cong., 3: 311-325.

Stern, E. M. and W. B. Stickle (1978), "Effect of Turbidity and Suspended Material in Aquatic Environments, Literature Review," Tech. Report D-78-21, U.S. Army Engineer Waterways Experiment Station, Vicksburg, Miss.

Sumeri, A. (1984), "Capped In-Water Disposal of Contaminated Dredged Material," Dredging and Dredged Material Disposal, Vol. 2, R. L. Montgomery and J. W. Leach, eds. (New York: American Society of Civil Engineers).

Taylor, J. L. and C. H. Saloman (1968), "Some Effects of Hydraulic Dredging and Coastal Development in Boca Ciega Bay, Florida," Fish. Bull., 767: 213-241.

Taylor, S. R. (1964), "Abundance of Chemical Elements in the Continental Crust: A New Table," Geochim. Cosmochim. Acta, 28: 1273-1285.

Thatcher, M. L. and D. R. F. Harleman (1972), "A Mathematical Model for the Prediction of Unsteady Salinity Intrusion in Estuaries," Report No. 144, Ralph M. Parsons Lab., Massachusetts Institute Technology, Cambridge, Mass.

Tramontano, J. M. and W. F. Bohlen (1984), "The Nutrient and Trace Metal Geochemistry of a Dredge Plume," Estuarine, Coastal and Shelf Science, 18: 385-401.

U.S. Army Corps of Engineers (1980), Publication Index and Retrieval System, Dredged Material Research Program, Technical Report DS-78-23 (Washington, D.C.: Government Printing Office).

U.S. Army Corps of Engineers (1979), "Reconnaissance Report: Dredged Material Containment in Long Island Sound," New England Division, Waltham, Mass.

Wilson, R. E. (1979), "A Model for the Estimation of the Concentrations and Spatial Extent of Suspended Sediment Plumes," Estuarine and Coastal Mar. Sci., 9: 65-78.

Wolman, M. G. and A. P. Schick (1967), "Effects of Construction on Fluvial Sediment, Urban and Suburban Areas of Maryland," Water Resource Res., 3: 451-464.

Zabawa, C. F. and J. R. Schubel (1974), "Geologic Effects of Tropical Storm Agnes on Upper Chesapeake Bay," Maritime Sediments, 10: 79-84.

SUMMARY OF COMMITTEE EXPERTISE

Don E. Kash, chairman, is George Lynn Cross Research Professor of Political Science and Research Fellow in Science and Public Policy at the University of Oklahoma. Dr. Kash was chief of the Conservation Division (now the Minerals Management Service) of the U.S. Geological Survey from 1978 to 1981. He has held appointments at the University of Missouri, Texas Technological University, Arizona State University, Purdue University, and Indiana University, and has served as a consultant to numerous governmental and private organizations.

John B. Herbich, vice chairman, is director of the Center for Dredging Studies at Texas A&M University, and professor of ocean and civil engineering. Dr. Herbich has served for many years as special consultant to the United Nations for research and education in ocean engineering and dredging, and has worked in several countries as a field, research, and consulting engineer. He has held appointments in coastal engineering at the University of Minnesota, Lehigh University, and the Hydraulic Institute of the University of Delft. He has served as director of the Western Dredging Association. Among Dr. Herbich's publications are Coastal and Deep Ocean Dredging, Offshore Pipelines - Design Elements, and Seafloor Scour, Design Guidelines for Ocean-Founded Structures.

J. W. Bean is president and chief executive officer of C. F. Bean Corporation, a dredging company, and registered professional engineer in the state of Louisiana. He was vice president of the International Association of Dredging Contractors, director of the World Dredging Association, and is a member and director of the National Association of Dredging Contractors.

W. Frank Bohlen is associate professor of marine science at the University of Connecticut. Dr. Bohlen's research has for the past several years concentrated on sediment transport processes, turbulence, and fluid mechanics, particularly in coastal waters and estuaries.

Allen B. Childress is the director of international coal and ore traffic for the Norfolk Southern Corporation, a position he has held since 1979. Before the merger of the Norfolk Western and Southern Railways into Norfolk Southern in 1982, Mr. Childress supervised coal transportation at the marine terminal in Norfolk for seven years having previously worked as trainmaster and supervisor at the railroad's inland terminals.

Richard L. Counselman, Jr., has been president of the Virginia Pilot Association for 13 years. Captain Counselman holds a First Class Pilot's License and Unlimited Master's License for Inland Waters. He is past president of the organization overseeing and operating Virginia International Terminals, and member of the Virginia Board of Commissioners to Examine Pilots.

J. Patrick Dowd is president of Coal Logistics Corporation, a topping-off service. He was vice president of the Investment Banking Division of Lehman Brohers, Kuhn Loeb, Inc., specializing in ship operations, coal properties, export terminals, and port and harbor projects, and served in a similar capacity with Smith, Barney, Harris, Upham and Company.

John S. Hollett, who served as a member of the Committee on National Dredging Issues through 1983, was director of the energy group of Crowley Maritime Corporation (an integrated shipping company) responsible for development of new business in dry bulk cargoes and petroleum. He previously served Crowley Maritime as commercial director of the international division and director of contract transportation in the Caribbean division.

Kenneth S. Kamlet is director of the Pollution and Toxic Substances Division of the National Wildlife Federation, a biochemical scientist, and member of the District of Columbia Bar. He was twice a member of the U.S. delegation to the London Dumping Convention, and has served on many policy review and planning committees. His environmental interests have for several years focused on the application of scientific and technical knowledge to policies addressing waste disposal and the handling of pollution and toxic substances.

Larry R. Olsen, who served as a member of the Committee on National Dredging Issues in 1984, was vice president for marketing of Crowley Maritime Corporation, having been director of coal transportation and managing director for Southeast Asia. Mr. Olsen worked for other ocean transportation companies in Canada and Belgium before joining Crowley Maritime and is now employed by a shipowning firm in Singapore.

Ernest L. Perry was executive director of the Port of Los Angeles for five years, retiring in late 1984. Dr. Perry served many years in the U.S. Army Corps of Engineers, directed heavy construction projects in the Middle East for international firms, and managed the Port of Tacoma for 11 years.

Clifford M. Sayre is director of logistics for E. I. Du Pont de Nemours and Company. Mr. Sayre has worked for Du Pont more than 30 years as a research engineer and supervisor, and in transportation--particularly the transportation of hazardous materials. He is a member of the Marine Board.

J. R. Schubel is dean and director of the Marine Sciences Research Center at the State University of New York at Stony Brook, and leading professor of oceanography. He served as vice president of the Estuarine Research Federation. Dr. Schubel's research interests are primarily in the geological and physical aspects of coastal sedimentation, particularly the processes that control the transportation and accumulation of fine-grained sediments in estuarine environments. He has also worked extensively on dredging and dredged material disposal problems in coastal and estuarine environments.

GENERAL DESIGN CRITERIA FOR
DREDGED NAVIGATIONAL FACILITIES

In 1972, the Permanent International Association of Navigation Congresses (PIANC) organized a commission to develop criteria for the reception of large vessels (IOTC, 1973). Six years later, a PIANC working group again studied the requirements of large vessels (200,000 DWT and greater), and published recommendations for port design (ICORELS, 1980), as did a working group of the International Association of Ports and Harbors (COLS, 1981). A review of PIANC guidelines is now under way, under the guidance of a new working group.

The recommendations of these international organizations include guidelines for the dimensions of channels and maneuvering areas, and also address forces of the physical environment, equipment, and training.

Maritime nations have developed general design criteria: those of Canada--TERMPOL--are based on prevention of oil pollution from marine casualties (Canadian Coast Guard, 1977). The general guidelines developed for ports and harbors in Japan (Bureau of Ports and Harbours, 1980) are detailed, reflecting the economic significance of ports to the country, its challenging natural environment, and the need to balance economical design and construction and safety margins.

The general design criteria used in the United States are developed by the U.S. Army Corps of Engineers. These were recently updated (U.S. Army Corps of Engineers, 1983).

Succeeding tables and figures describe and compare these general criteria. It should be understood that all these sets of criteria acknowledge the importance of "(1) the several site-specific factors of great importance to design, (2) the need for consultations with shipowners, pilots, and others, and (3) the need to employ analysis and design tools" (Crane, 1983).

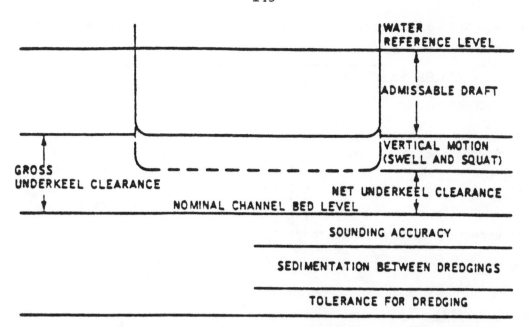

(a) Conventional Net Underkeel Clearance Calculation, Definitions from PIANC (Permanent International Association of Navigation Congresses)

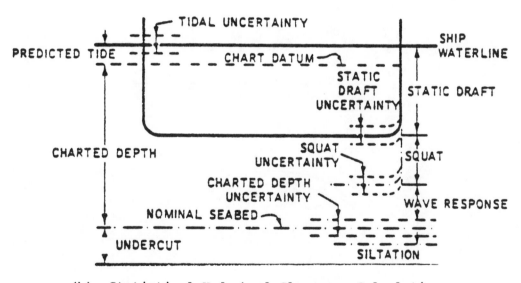

(b) Statistical Underkeel Clearance Calculation

FIGURE B-1
SOURCE: C. Lincoln Crane, Jr.

146

TABLE B-1 General Criteria for Depths of Dredged Navigational
Facilities

<u>U.S. Army Corps of Engineers</u>
    Ship's draft and sinkage + allowances for wind, waves, currents,
type of bottom, etc.

<u>PIANC</u>
    Gross underkeel clearance   1.20 x ship's draft, exposed
                                          1.15 x ship's draft, waiting area,
                                        exposed
                                        1.07 x ship's draft, calmest area,
                                        least ship speed (berthing)
    Net underkeel clearance at least 0.5 m (1.7 ft)

<u>SHIPOWNERS</u>
    Statistical

<u>TERMPOL (CANADA)</u>
    1.15 x ship's draft, exceptions require special underkeel clearance
survey

<u>Japan</u>
    Depth of maneuvering basin (1.10 x ship's draft) + allowances for
wind, waves, currents, type of bottom, etc.

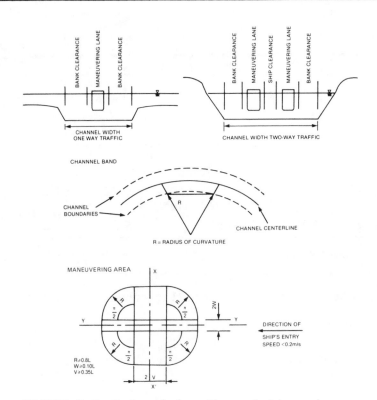

FIGURE B-2 Determining Channel Dimensions

TABLE B-2 Comparison of Port Design Guidelines for Channel Width (Referred to Dimensions of Design Vessel(s))

	One-Way Traffic			Two-way traffic			Definitions and Comments
	COE*[1]	TERMPOL[2]	PIANC[3]	COE*[1]	TERMPOL[2]	PIANC[3]	
Maneuvering Lane Straight channel Bends of 26° 40°	2.0 beam 4.2 beam 4.9 beam	2.0 beam 4.0 beam	Additional Width $\Delta W = L^2/BR+$ $= length^2/(beam \times radius) +$	2.0 beam 4.2 beam 4.9 beam	2.0 beam 4.0 beam		Maneuvering lane (ML): - Lane in which a single vessel maneuvers - Accounts for uncertainty in vessel position and time lag to correct position - Applies to straight and turning, not additive - Required for each ship (i.e., MLX2 for 2-way traffic) - Can reduce if operational limits applied (e.g., 2 design ships don't pass each other and no passing in turns)
Bank Clearance Each side	.6 beam +	1.0 beam	Safety margin .5	.6 beam +	1.0 beam		Bank Clearance (BC): - Clearance to avoid bank suction - Required on both sides, BC x 2 for all channels
Passing Ship Clearance Between Lanes				.8 beam	1.0 beam		Passing Ship Clearance (PSC): - Between MLs to avoid interference
Total Width w/o Weather, Current Straight Channels Bends of 26° 40°	3.2 beam 5.4 beam 6.1 beam +	4.0 beam 6.0 beam		6.0 beam + 10.4 beam 11.8 beam	7.0 beam 11.0 beam		Total straight Channel Width Without Current Clearance (W w/o WCC) - Artificial since cannot usually be used without WCC, even for mild conditions
Weather and Current Clearance	.9 beam Each side	1.0 beam Total both sides		(.2+.9x2) beam Total both sides	1.0 beam Total per ship lane		Weather and Current Clearance (WCC): - For beam wind, current, and waves which cause a yaw angle (10-15° max.) - Additional clearance should be made for varying conditions (gusts) and other factors - For design, must correspond to limiting environmental operating condition
Total Width Straight Channels Bends, of 26° 40°	5.0 beam 7.2 beam 7.9 beam +	5.0 beam 7.0 beam	5 beam 5 beam + length²/ (beam x radius)	8.0 Beam + (2) 12.4 beam + 13.8 beam	9.0 beam 13.08 beam	8 beam	Total Width of Channel (W) at Design Depth: - In general, minimums are shown for design, however, local conditions must be considered
Parallel Bends Radius for Bends, of <25° 26°-35° >36° Width transition (total)	 >5 length 1:10	w/o tugs 3 length 5 length 8 length 1:10	 5-10 length	SAME AS ONE WAY			Radius of Turn (R): - Radius of the channel centerline for bends - TERMPOL gives max rudder angle (δ) of 15° (related to the design ships' turning diameters at δ = 15°) Transition: The ratio of widening (sum of both sides) to length along channel
Approximate Ship Size **	Small/medium size tanker	Not specified	Very ultra large crude carrier	Small/medium size tanker	Not specified	Very/ ultra large crude carrier	Ship Size: - The approximate ship size for which the specific guide is intended (in terms of tankers)
General Equation	Width = maneuvering lane + 2 bank clearance + weather & current clearance			Width = 2 maneuvering lane + 2 bank clearance + passing ship clearance + weather and current clearance			General equation relating the individual components to the total width

*Using values for "poor vessel controllability"
[1] U.S. Army Corps of Engineers (1983).
[2] Canadian Coast Guard (1977).
[3] Permanent International Association of Navigation Congresses (1980).

**Small-size Tanker: 16,000 - 60,000 DWT
Medium-size Tanker: 60,000 -120,000
Very large crude carrier: 120,000 -320,000
Ultra large crude carrier: more than 320,000 DWT

TABLE B-3 General Criteria for Turning Basins and Anchorages

TURNING BASINS
U.S. Army Corps of Engineers
Equal to area of circle with radius = 1.5 x ship's length over-all + allowances for congestion, sedimentation, current, etc.

Side parallel to channel longer, ends angled 45° to channel boundary

PIANC
Equal to circular area with diameter = ship's length over-all

Elliptical shape recommended

Japan
Equal to circular area with radius = 1.5 x ship's length over-all,

ANCHORAGES
U.S. Army Corps of Engineers
Free-swinging

Area = area of circle with radius = ship's length over-all + anchor chain (5x to 6x water depth)

Fixed dolphins, berths

Width = 1.5 x ship's beam
parallel to channel

PIANC
None

Japan

Design Objective	Mooring	Seabed/Wind	Radius (LAO = ship's length over-all)
Offshore; waiting	Swinging	Good anchoring	LOA + (6 x water depth)
		Bad anchoring	LOA + (6 x water depth) + 30 m (99 ft)
Mooring in storm	Mooring with 2 anchors	Good anchoring	LOA + (4.5 x water depth)
		Bad anchoring	LOA + (4.5 x water depth) + 25 m (82.5)
		Wind vel. - 20 m/sec (40 km)	LOA + (3 x water depth) + 90 m (297 ft)
		Wind vel. = 30 m/sec (60 km)	LOA + (4 x water depth) + 145 m (478.5 ft)

APPENDIX C

QUESTIONNAIRE TO PILOTS'
ORGANIZATIONS*

1. Name of pilot organization

2. What waterways and ports are served by your organization?

3. What are the various channel sizes (depth/width) encountered along this route?

4. What are the types and sizes of vessels transiting these channels?

5. What major commodities are carried by these vessels?

6. Are the channels adequate for the maneuvering requirements of the vessels you pilot through them?

7. What vessels present the greatest problems in maneuvering and control?

8. What areas pose navigational difficulties in the entrances, channels, and harbor turning basins of your pilotage route?

9. In the areas described above, what major factors contribute to controllability and maneuvering problems (e.g., channel constriction, shoaling or underkeel clearance, high winds, strong currents)?

10. What practices have pilots agreed to among themselves to compensate for the deficiencies in channel design (one-way traffic, restricted passing/overtaking in bends and turns, etc.)?

11. Would deepening or widening entrance channels, river or approach channels, and turning basins improve ship maneuverability and control?

---

*Distributed, collected, and analyzed by the Technical Panel on Ports, Harbors, and Navigational Channels, Study of National Dredging Issues.

12. What specific areas should be deepened and/or widened to improve vessel maneuverability and control in your pilotage area?

13. Do your pilots consider the navigation aids in your area adequate or in need of improvement?

14. Would improved navigational aids substitute for improvements in channel depth, width, or design?

15. What other methods are used in your pilotage area to compensate for inadequate channels (e.g., use of tugboats, high-water transit, transit with or against current)?

16. What is the maximum draft of vessels calling at your ports?

17. What underkeel clearance is recommended for transiting your channels?

18. What is the type of bottom in areas of critical underkeel clearance (sand, rock, variable)?

19. How was the recommended underkeel clearance established?

20. What traffic control systems (other than those operated by government agencies) have been established to compensate for deficiencies in the navigational channels of your area?

21. How often is maintenance dredging performed in your channels?

22. Do your pilots consider the maintenance dredging schedule adequate for the channels along your pilotage route?

23. Generally, what is your assessment of the adequacy of your waterways for the traffic and conditions experienced by your pilots and what solutions would you recommend for improvement?

APPENDIX *D*

<u>REQUEST FOR INFORMATION FROM
PORTS OF OTHER MARITIME NATIONS</u>*

Foreign ports were asked to supply the following information:

1. Depth and width of navigational channels

2. Depth in sheltered areas and in the harbor

3. Maximum draft of vessels allowed to transit the port

4. Port use by size of ships (port calls per deadweight-ton categories)

5. Annual tonnage

6. How many larger vessels are excluded, or transit partially loaded?

7. Are operational practices employed owing to channel (or other) limitations?

8. Does the port plan to expand capacity? If so, what are the port's plans for:

    a.   Deepening, widening (or both) of navigational channels by dredging

    b.   Offshore lightering/topping-off

    c.   New deep-water ports

    d.   Reception of broad-beam vessels

---

*Distributed, collected, and analyzed by the Committee on National Dredging Issues (59 ports responding).

APPENDIX *E*

POLICY AND LEGISLATION PERTINENT TO DREDGING

(a) Public laws.
  (1) American Folklife Preservation Act, Pub. L. 94-201; 20 U.S.C. 2101, et seq.
  (2) Anadromous Fish Conservation Act, Pub. L. 89-304; 16 U.S.C. 757, et seq.
  (3) Antiquities Act of 1906, Pub. L. 59-209; 16 U.S.C. 431, et seq.
  (4) Archeological and Historic Preservation Act, Pub. L. 93-291; 16 U.S.C. 469, et seq. (Also known as the Reservoir Salvage Act of 1960, as amended; Public Law 93-291, as amended; the Moss-Bennett Act; and the Preservation of Historic and Archeological Data Act of 1974.)
  (5) Bald Eagle Act; 16 U.S.C. 666.
  (6) Clean Air Act, as amended, Pub. L. 91-604; 42 U.S.C. 1857h-7, et seq.
  (7) Clean Water Act, Pub. L. 92-500; 33 U.S.C. 1251, et seq. (Also known as the Federal Water Pollution Control Act; and Public Law 92-500, as amended.)
  (8) Coastal Zone Management Act of 1972, as amended, Pub. L. 92-583; 16 U.S.C. 1451, et seq.
  (9) Endangered Species Act of 1973, as amended, Pub. L. 93-205; 16 U.S.C. 1531, et seq.
  (10) Estuary Protection Act, Pub. L. 90-454; 16 U.S.C. 1221, et seq.
  (11) Federal Environmental Pesticide Control Act, Pub. L. 92-516; 7 U.S.C. 136.
  (12) Federal Water Project Recreation Act, as amended, Pub. L. 89-72; 16 U.S.C. 460-1(12), et seq.
  (13) Fish and Wildlife Coordination Act of 1958, as amended, Pub. L. 85-624; 16 U.S.C. 661, et seq. (Also known as the Coordination Act.)
  (14) Historic Sites of 1935, as amended, Pub. L. 74-292; 16 U.S.C. 461, et seq.
  (15) Land and Water Conservation Fund Act, Pub. L. 88-578; 16 U.S.C. 4601-4601-11, et seq.
  (16) Marine Mammal Protection Act of 1972, Pub. L. 92-522; 16 U.S.C. 1361, et seq.

152

(17) Marine Protection, Research and Sanctuaries Act of 1972, Pub. L. 92-532; 33 U.S.C. 1401, et seq.

(18) Migratory Bird Conservation Act of 1928; 16 U.S.C. 715.

(19) Migatory Bird Treaty Act of 1918; 16 U.S.C. 703, et seq.

(20) National Environmental Policy Act of 1969, as amended, Pub. L. 91-190; 42 U.S.C. 4321, et seq. (Also known as NEPA; often incorrectly cited as the National Environmental Protection Act.)

(21) National Historic Preservation Act of 1966, as amended, Pub. L. 89-655; 16 U.S.C. 470a, et seq.

(22) Native American Religious Freedom Act, Pub. L. 95-341; 42 U.S.C. 1996, et seq.

(23) Resource Conservation and Recovery Act of 1976; Pub. L. 94-580; 7 U.S.C. 1010, et seq.

(24) River and Harbor Act of 1899, 33 U.S.C. 403, et seq. (Also known as the Refuse Act of 1899.)

(25) Submerged Lands Act of 1953, Pub. L. 82-3167; 43 U.S.C. 1301, et seq.

(26) Surface Mining Control and Reclamation Act of 1977, Pub. L. 95-89; 30 U.S.C. 1201, et seq.

(27) Toxic Substances Control Act, Pub. L. 94-469; 15 U.S.C. 2601, et seq.

(28) Watershed Protection and Flood Prevention Act, as amended, Pub. L. 83-566; 16 U.S.C. 1001, et seq.

(29) Wild and Scenic Rivers Act, as amended, Pub. L. 90-542; 16 U.S.C. 1271, et seq.

(b) Executive orders.

(1) Executive Order, 11593, Protection and Enhancement of the Cultural Environment, May 13, 1979 (36 FR 8921; May 15, 1971).

(2) Executive Order, 11988, Floodplain Management, May 24, 1977 (42 FR 26951; May 25, 1977).

(3) Executive Order, 11990, Protection of Wetlands, May 24, 1977 (42 FR 26961; May 25, 1977).

(4) Executive Order, 11514, Protection and Enhancement of Environmental Quality, March 5, 1970, as amended by Executive Order, 11991, May 24, 1977.

(5) Executive Order, 12088, Federal Compliance with Pollution Control Standards, October 13, 1978.

(c) Other Federal policies.

(1) Council on Environmental Quality Memorandum of August 11, 1980: Analysis of Impacts on Prime or Unique Agricultural Lands in Implementing the National Environmental Policy Act.

(2) Council on Environmental Quality Memorandum of August 10, 1980: Interagency Consultation to Avoid or Mitigate Adverse Effects on Rivers in the Nationwide Inventory.

(3) Migratory Bird Treaties and other international agreements listed in the Endangered Species Act of 1973, as amended, Section 2(a)(4).

(d) Selected state legislation, lead agencies, and concerns
   (1) Maine: Department of Environmental Protection for coastal
       and great ponds projects (Tidal Wetlands Act 38 Maine Revised
       Statutes Annotated Sections 471-478 and Great Ponds Act 38
       MRSA Sections 386-396, respectively). Department of Inland
       Fish and Wildlife for fill projects on rivers and streams
       (Alteration of Rivers, Streams and Brooks Act 12 MRSA
       Sections 7776-7780). The Board of Environmental Protection
       may establish any reasonable requirement to ensure that the
       applicant does not contravene environmental quality.

   (2) New Hampshire: Water Supply and Pollution Control Commission
       (Resource Statutes Annotated, Subsection 149.8A) and the
       Wetlands Board (RSA, Subsection 483A). The Water Supply and
       Pollution Control Commission requires that there be no
       degradation of water quality.

   (3) Massachusetts: Conservation Commission of locality directly
       affected by the project (State Wetlands Protection Law,
       Chapter 131, Section 40). A local Conservation Commission
       may attach special conditions to an application to ensure
       proper response to its concerns when discharge to a wetlands
       is proposed.

   (4) Rhode Island: Coastal Resources Management Council.
       (General Laws, Chapter 279, Section 1). The Coastal
       Resources Management Council is concerned with state coastal
       plan consistency and permitting activities in territorial
       waters and saltwater wetlands.

   (5) Connecticut: Commissioner of the Department of Environmental
       Protection, (Marine Mining Statute, Section 25-7d for new
       dredging work and structures and Dredging Statute, Section
       25-11 for regulating building of marina structures). The
       Department of Environmental Protection requires containment
       of materials disposed of on upland sites. In-water disposal
       permits may require special conditions to protect fish and
       wildlife recommended by the U.S. Fish and Wildlife Service
       and/or the National Marine Fisheries Service.

   (6) New York Department of Environmental Conservation
       (Environmental Conservation Law, articles). The Department
       of Environmental Conservation may specify seasonal
       restrictions to protect spawning. It may also specify
       certain types of dredging and containment procedures to
       alleviate environmental impact.

155

(7) California: California Coastal Commission (Proposition 20, 1972; California Coastal Zone Act, 1976). Requires port master plan; lead agency for review of port projects. Water Control Board (California Resources Code). Permit authority over effects of dredging/filling on water quality. Department of Fish and Game (California Resources Code). Review and comment authority over effects of proposed projects on fish and wildlife. Air Resources Board (California Resources Code). Permit review authority over sources of stationary (point-source) air pollution has been applied to dredging equipment and port facilities.

(8) Oregon: Department of Land Conservation and Development. Statewide goals and guidelines for coastal resources. Department of Fish and Wildlife (Oregon Administrative Rule). Classifies estuaries.

(9) Washington: Shoreline Hearings Board (Shoreline Management Act). Permit appeal authority. Department of Ecology (Washington Resources Code). Water and air quality permit authority; review of proposed projects for effects on fish and wildlife.

APPENDIX *F*

COMMISSIONED PAPERS AND BACKGROUND MATERIALS
PREPARED FOR STUDY OF NATIONAL DREDGING ISSUES*

Feldman, R. and E. Haber, Bibliography of Selected National Dredging Issues, 2 vol. (Bibliography, 612 pp).

Gerwick, B. C., Alternatives to Dredging (Commissioned Paper, 21 pp).

Gunn, B., Money and Ports: User Fees and Cost Sharing (Background Paper, 51 pp).

Krone, R. B., Minimizing the Cost of Maintaining Navigable Water Depths in Estuaries (Commissioned Paper, 30 pp).

McSweeny, J. and E. Margolin, Analysis of the Impact and Incidence of Alternate Deep Water Port Cost Recovery Mechanisms (Commissioned Paper, 30 pp).

Oppenheimer, C. H., Environmental Effects of Dredged Material (Commissioned Paper, 97 pp).

Record of the Public Meeting of the Study of National Dredging Issues, September 29, 1983, Washington, D.C. (Transcript and Formal Submissions).

Report of the Technical Panel on Ports, Harbors, and Navigational Channels (100 pp).

---

*Single copies available on request from the Marine Board, National Research Council, 2101 Constitution Ave., N.W. Washington, DC 20418.

APPENDIX G

FIGURES AND TABLES

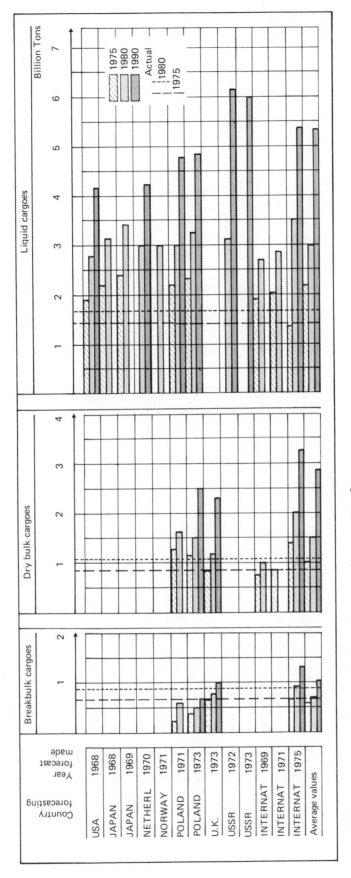

FIGURE 1 Selected Forecasts of Oceanborne Trade

160

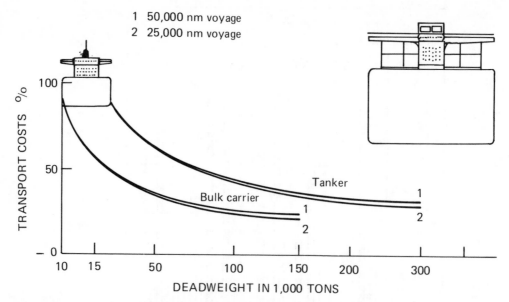

FIGURE 2a Relationship of Tanker and Bulk Carrier Vessel Size to
Transport Costs per Ton of Cargo
SOURCE: Schonknecht et al., 1983

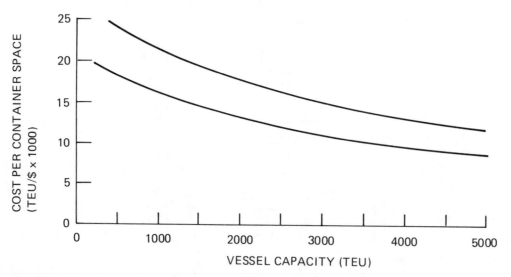

FIGURE 2b Relationship of Container-Carrying Capacity to
Cost/Container
SOURCE: C. R. Cushing, 1984.

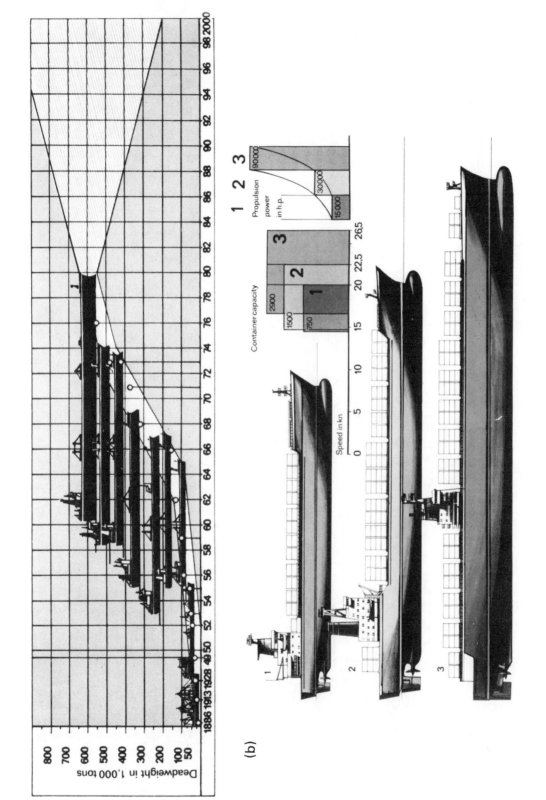

(a)

(b)

FIGURE 3 Worldwide Trend Toward Larger Vessels (a) Tankers (b) Containerships
SOURCE: Schonknecht et al., 1983, Ships and Shipping of Tomorrow (Centreville,
Md.: Cornell Maritime Press)

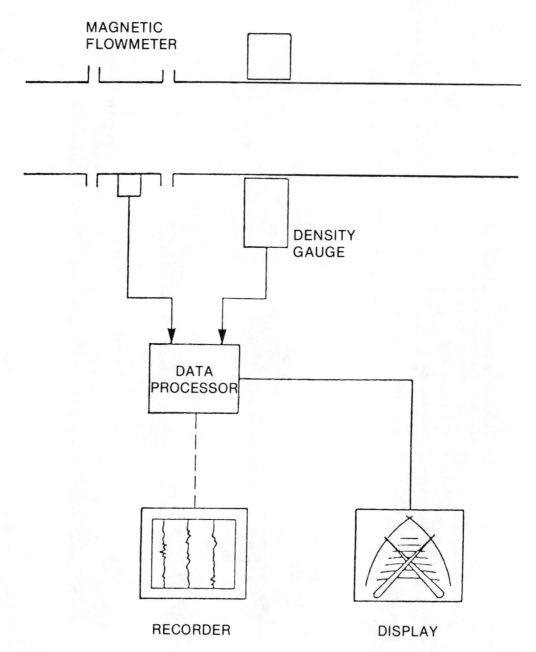

FIGURE 4 A Production Meter System With a Nucleonic Density Gauge and a Crossed-Pointer Display

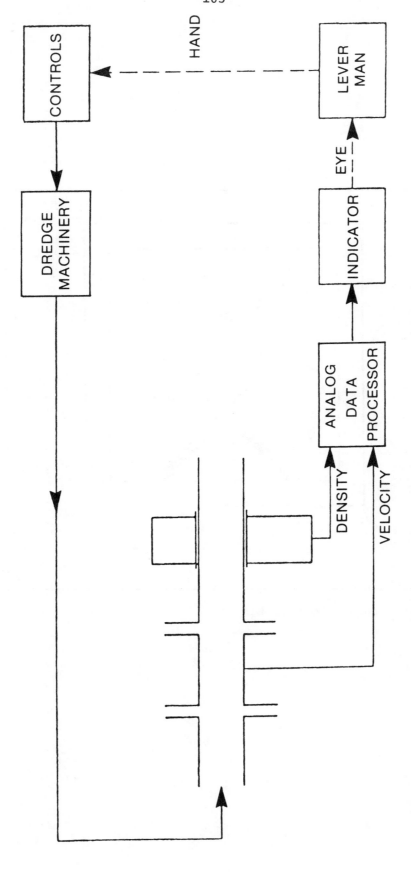

FIGURE 5   The Leverman and a Production Meter System as Parts of a Closed-Loop Control System

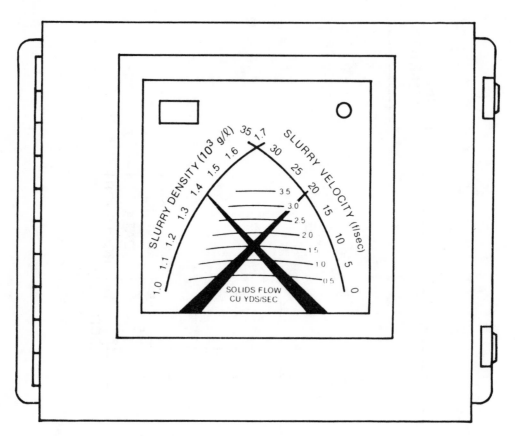

FIGURE 6   Crossed Pointer Display

165

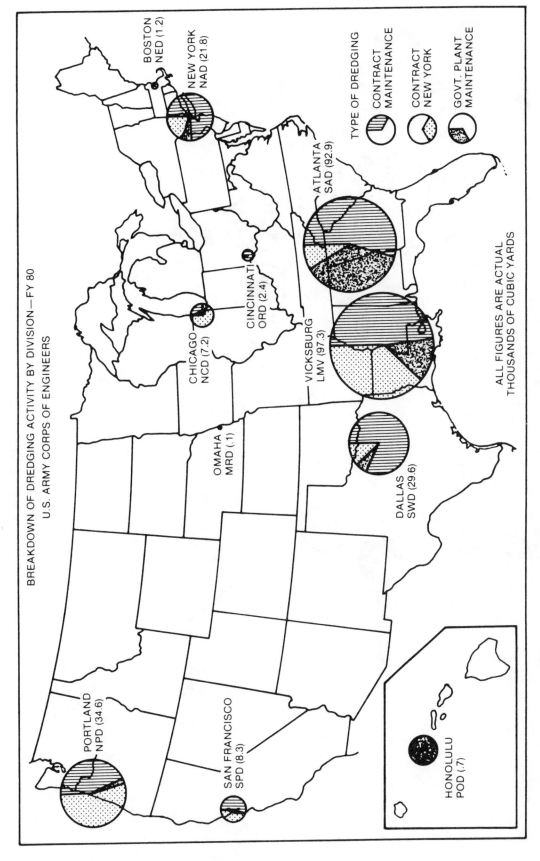

FIGURE 7 Dredging Program of the U.S. Army Corps of Engineers (FY80)

166

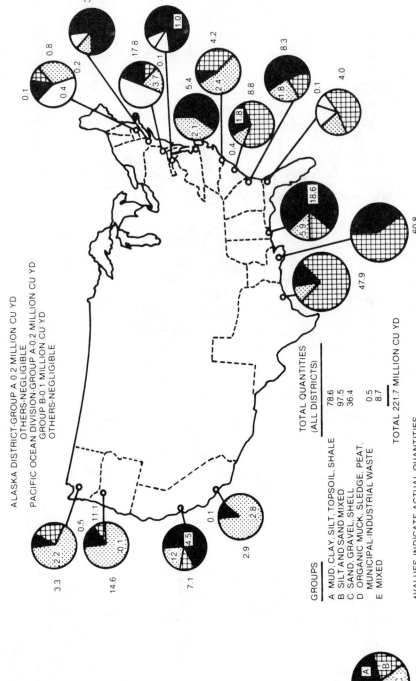

GROUPS

A MUD, CLAY, SILT, TOPSOIL, SHALE
B SILT AND SAND MIXED
C SAND, GRAVEL, SHELL
D ORGANIC MUCK, SLEDGE, PEAT.
   MUNICIPAL-INDUSTRIAL WASTE
E MIXED

TOTAL QUANTITIES (ALL DISTRICTS)
78.6
97.5
36.4
0.5
8.7

TOTAL 221.7 MILLION CU YD

†VALUES INDICATE ACTUAL QUANTITIES
 IN MILLIONS OF CU YD

TOTAL DREDGING BY CE
DISTRICT (CU YD)

25 TO 60 MILLION

0 TO 25 MILLION

ALASKA DISTRICT-GROUP A 0.2 MILLION CU YD
OTHERS-NEGLIGIBLE
PACIFIC OCEAN DIVISION-GROUP A-0.2 MILLION CU YD
GROUP B-0.1 MILLION CU YD
OTHERS-NEGLIGIBLE

FIGURE 8   Characterization of Materials Dredged by The U.S. Army Corps of Engineers
SOURCE: From Pequegnat, et al., 1978

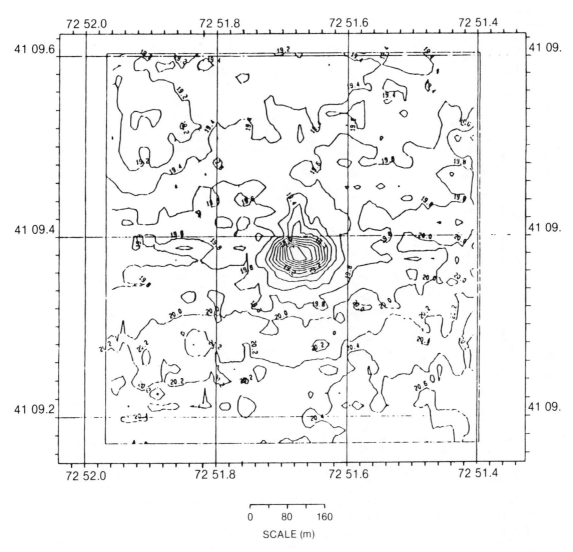

FIGURE 9 An Example of a Precision Dumping Operation at the Central Long Island Sound Disposal Site (1983)

168

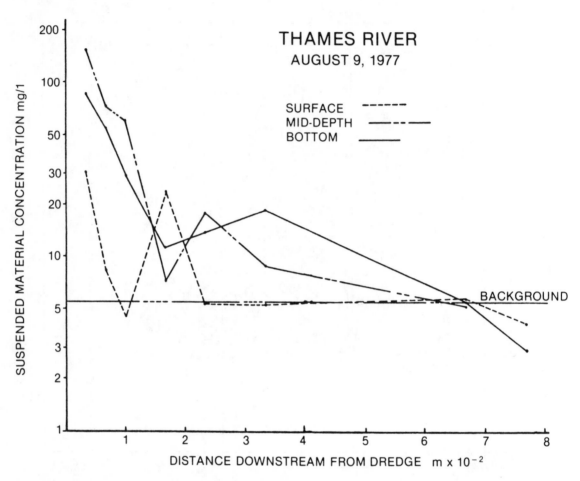

FIGURE 10 Suspended Material Concentrations in the Wake of a
Mechanical Dredging Operation
SOURCE: Bohlen, et al.

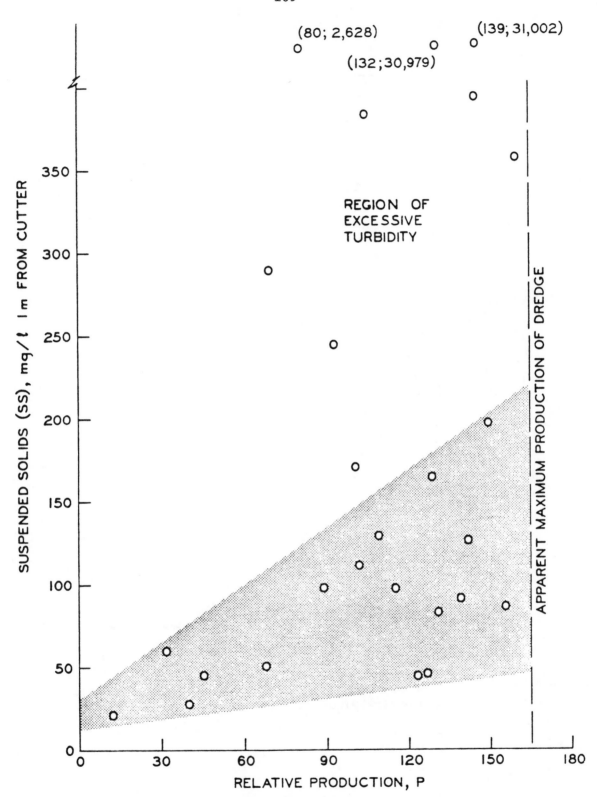

FIGURE 11 Relationship Between the Concentration of Suspended Solids
1 m From the Cutter and the Relative Production of a 61 cm (24 in.)
Cutterhead Dredge
SOURCE: Barnard, 1978

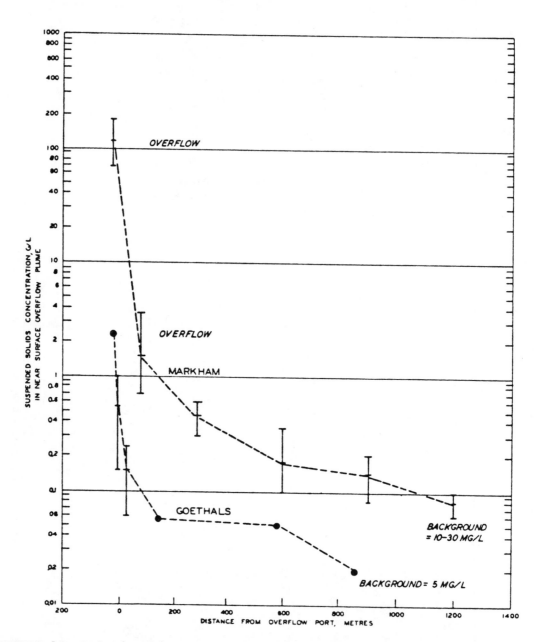

FIGURE 12   Relationship Between Concentration of Suspended Solids in
the Near-Surface Overflow Plume and the Distance (in m) Downstream of
the Overflow Ports
SOURCE: Barnard, 1978

171

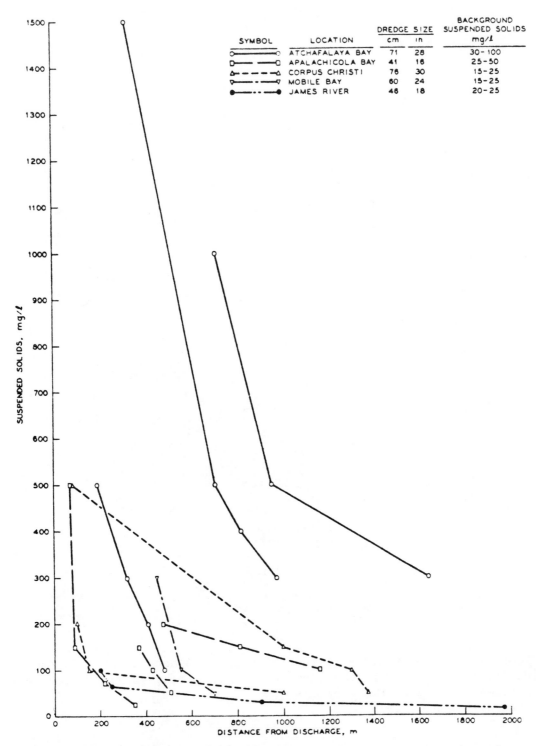

FIGURE 13 Relationship Between Suspended Solids Concentration Along
the Plume Centerline and Distance Downcurrent From Several Open-Water
Pipeline Disposal Operations
SOURCE: Barnard, 1978

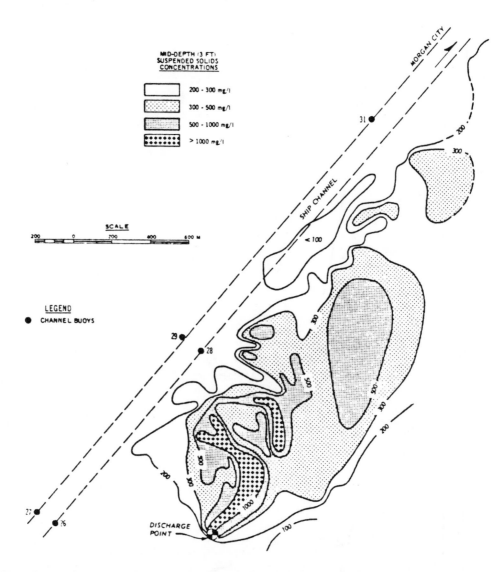

FIGURE 14 Mid-Depth (0.9 m) Turbidity Plume Generated by a 71 cm (28 in.) Pipeline Disposal Operation in the Atchafalaya Bay. Current Flow is Generally Toward the Northeast
SOURCE: Barnard, 1978

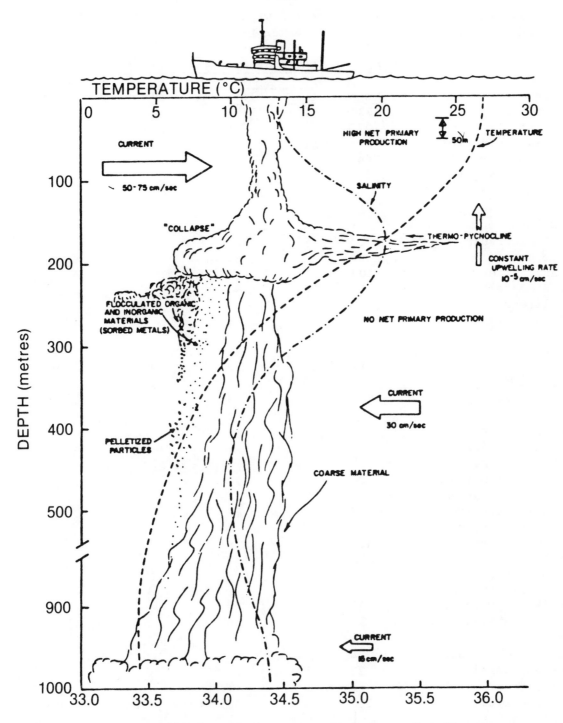

FIGURE 15 Characteristics of the Descending Mass of Sediment
Discharge From a Surface Barge
SOURCE: Pequegat et al., 1978

TABLE 1 Number of Vessels in Foreign Trade Calling on Ports of the United States, 1980, by Type and Design Draft

Vessel Type	Total	Number of Vessels by Design Draft ft					
		1-15	16-45	46-50	51-55	56+	Other
Freighter	1730	31	1698	1			
Tanker	1077	3	742	123	96	113	
Freighter/refrig.	329	11	318				
Bulk carrier	2642		2532	57	44	9	
Combo. pass. & cargo	85	5	80				
Combo/refrig.	4		4				
Ore/oil carrier	54		29	13	33	9	
Whaling tanker	1		1				
Containership	322	7	325				
Ore carrier	84		81	2	1		
Car carrier	103		102			1	
LPG tanker	95	2	93				
Colliers	7		7				
Asphalt tanker	9		9				
Bitumen	5		5				
Chemical tanker	193	1	192				
LNG tanker	12		12				
Molasses tanker	6		6				
Phosphorus tanker	3		3				
Sulphur tanker	6		6				
Wine tanker	1		1				
Barge carrier	1		1				
Cattle carrier	2		2				
Container/barge car.	20		20				
Container/car car.	1		1				
Container/Ro-Ro	16	4	12				
Pallet carrier	6		6				
Partial container	421	5	416				
Roll-on/Roll-off	153	14	139				
Timber carrier	6		6				
Bauxite carrier	1		1				
Bulk car carrier	98		97				
Bulk/containership	30		30				
Bulk/oil	1		1				
Bulk/timber car.	1		1				
Cement carrier	22	1	21				
Limestone	1		1				
Ore/bulk/oil	180		48	61	30	41	
Salt carrier	1		1				
Wood chip carrier	5		55				
Barge (dry cargo - domes)		5					5
Barge (tanker -domestic)		3					3
Total	7802	84	7105	258	174	173	8

SOURCE: Maritime Administration, Office of Port and Intermodal Development.

TABLE 2 Proposed Deep-Draft Port Dredging Projects*

Project/ Port	Proposed New Dimensions	Design Basis
Baltimore Harbor and channels, Baltimore, Maryland	50' depth	Bulk carrier: 140,000 DWT, 950' length, 141' beam, 55' draft (light-loaded to less than 55')
Hampton Roads: Norfolk, Newport News, Virginia	55' depth, existing channels; possibly new 57' depth segment in Atlantic	Increments below existing depths analyzed, design vessel selected to fit. Design basis being refined through vessel simulations Current design vessel: Bulk carrier (coal), 120,000 DWT, 901' length, 133' beam, 52' draft
Mississippi River, Gulf of Mexico to New Orleans and Baton Rouge, Louisiana	55' depth, 750' width	Optimal net benefits for various channel depth-increments--vessel selected to fit. For depth: 122,000 DWT, 905' length, 132' beam, 51' draft For width: 105,000 DWT, 880' length, 134' beam, 51' draft
	turning basin, 1360' x 4000'	For turning basin: 122,000 DWT, 905' length, 132' beam, 51' draft
Mobile Harbor, Alabama	57' depth, bar channel 55' depth, 550' width, main channel	150,000 DWT, 953' length, 142' beam, 57' draft, assuming light-loading by up to 5' of draft

Proposed Deepening/Widening for Latest-Generation Containerships

Project/ Port	Proposed New Dimensions	Design Basis
Gowanus Creek Channel, New York/New Jersey	45' depth	Existing vessels using channel: 880' length, 106' beam, 41' draft
Kill van Kull New York/New Jersey	45' depth	Containership: 880' length, 106' beam, 41' draft
Charleston, South Carolina	40' depth, 600' width; 1200' turning basin	For width, containership: 800' length, 110' beam, 40' draft For depth, 47,000 DWT tanker: 800' length, 105' beam, 38' draft

TABLE 2   (continued)

Project/ Port	Proposed New Dimensions	Design Basis
Savannah, Georgia	500' width (section of channel); widening turning basin from 400' to 500'	Containership: 863' length, 105' beam, 38' draft
Oakland, California	42' depth, 800' width, outer harbor; eliminate dogleg; add turning basin, 1800' diameter	Existing and expected containerships: Panamax length and width selected for design basis: 950' length, 105' beam, 28' to 43' draft
Richmond, California	(same as for Oakland)	

Other Deepening Proposals 35 ft to 45 ft

Elizabeth River, Norfolk, Virginia	45' depth	Vessels carrying grain, residual fuel: 60,000 DWT, 729' length, 104' beam, 42' draft
Elizabeth River, South Branch, Norfolk, Virginia	40' depth	Vessels carrying fertilizer, grain, residual fuel: 37,000 DWT, 660' length, 90' beam, 37' draft
Northwest Branch, East Channel, Baltimore, Maryland	49' depth	Bulk carrier: 100,000 DWT, 850' length, 124' beam, 49' draft; tanker: 80,000 DWT, 811' length, 122' beam, 43.6' draft
Northwest Branch, West Channel, Baltimore, Maryland	40' depth	Bulk carrier: 100,000 DWT, 850' length, 124' beam, 49' draft (grain) Bulk carrier: 40,000 DWT, 650' length, 91' beam, 37' draft (sugar)
New Haven Harbor, Connecticut	40' depth, 500' width, main channel, realignment of one segment, widening bend from 560' to 780', 1200' turning basin	Coastwise tanker: 62,000 DWT, 780' length, 110' beam, 42' draft (assuming use of tide and light-loading)--design vessel selected to fit maximized net benefits at 40' depth

177

TABLE 2 (continued)

Project/ Port	Proposed New Dimensions	Design Basis
Blair Waterway, Tacoma, Washington	45' depth, new turning basin, 1200'	Existing vessels using waterway: 90,000 DWT, 820' length, 122' beam, 46' draft
	41' depth upper reach	44,000 DWT, 658' length, 201' beam, 37' draft
Sitcum Waterway, Tacoma, Washington	40' depth outer channel	65,000 DWT, 850' length, 115' beam, 40' draft
	35' depth, inner Sitcum Waterway	25,000 DWT, 661' length, 87' beam, 33' draft
Wilmington Harbor, North Carolina	500' width, jetty entrance channel	New benefits optimized at 35'; various design vessels tested--Tanker: 25,000 DWT, 585' length, 80' beam, 32' draft; Bulk carrier: 21,000 DWT, 560' length, 74' beam, 32' draft
	35' depth, 900' x 1000' turning basin	For widths--47,000 DWT 736' length, 99' beam, 38' draft
		For turning basin-- 36,000 DWT, 660' length, 91' beam, 35' draft
St. Thomas Harbor, Virgin Islands	38' depth, 500' width, new turning basin, 1200' x 1600'	Cruise ships: 10,800 DWT, 760' length, 90' beam, 32' draft
		For width: 4,800 DWT, 640' length, 80' beam, 28' draft
Sacramento River, Sacramento, California	35' depth	Summary design vessel based on combination of broader-beam wood-chip vessel, and deeper-draft dry bulk carrier able to use channel if deepened: 20,000 DWT, 520' length, 83' beam, 32' draft
John F. Baldwin and Stockton Channels, Stockton, California	35' depth, widening one reach and turns	Summary design vessel: 23,000 DWT, 575' length, 75' beam, 32' draft

TABLE 2 (continued)

Project/ Port	Proposed New Dimensions	Design Basis
Freeport, Texas	47' depth, 400' width, entrance channel, realign- ment and extension 45' depth, 400'' width, jetty channel, relocation of north jetty	Tanker: 63,000 DWT, 800' length, 110' beam, 41' draft
Brazos Island Harbor, Brownsville Texas	44' depth, 400' width, extension of north jetty 42' depth, 300' width, main channel 42' depth, 1200' turning basin	Tanker: 43,000 DWT, 665' length, 93' beam, 38' draft (+ 56' barge lane in turning basin)
Corpus Christi, Texas	47' depth, outer bar channel 45' depth, remain- ing waterway	Bulk carrier: 75,000 DWT, 800' length, 113' beam, 41' draft (grain)

aProposed projects having reached approval levels of Chief of Engineers or beyond.

*SOURCE: U.S. Army Corps of Engineers, Planning Division, Directorate of Civil Works.

TABLE 3 Draft Of Vessels Sailing To Or From Selected U.S. Ports - 1981

Port	Auth. Depth	30	31	32	33	34	35	36	37	38	39	40	41	42	43	44	45	46	47	48	49	50-60
Anchorage, AK	36							2														
Anacortes, WA	33(a)			16	18	54	73															
Bellingham, WA	30	5	24																			
Everett, WA	30	25	10	16	11	36																
Grays Harbor, WA	30	12	9	23	19	71	1															
Kalama, WA	40(b)								7	0	1											
Longview, WA	40(b)								18	5	4	1										
Olympia, WA	30	21																				
Seattle, WA	34				112	69		23	98													
Tacoma, WA	35	49	36	35	33	39	163															
Vancouver, WA	40																					
Astoria, OR	40(b)								4	1												
Coos Bay, OR	35					25	15	5	2													
Portland, OR	40(b)								29	22	0	1										
Yaquina Bay, OR	40(c)																					
Humboldt, CA	35					8	6	4														
San Francisco, CA	40										2											
Redwood City, CA	30	6	1	1																		
Oakland, CA	35					129	89	28	19	9	8											
Richmond, CA	35					66	71	46	28	11	8	3	0	1	3	5						
San Pablo, CA	35					300	89	36	14	7	6	3	3	2	1	3		4	1		1	1
Stockton, CA	30																					
Sacramento, CA	30	76																				
Los Angeles, CA	45(d)															16	5	2		2	5	13
Long Beach, CA	45(e)															28	21	16	4	12	11	100
San Diego, CA	35					16	6	6														
Searsport, ME	35					6	5	5														
Portland, ME	45															10	23	1				
Portsmouth, NH	35																					
Salem, MA	32																					
Weymouth, MA	35																					
Boston, MA	40										8	3										
Dorchester, MA	35																					
Fairhaven, MA	39																					
Fall River, MA	35			8	32																	
New Bedford, MA	30																					
Providence, RI	40										1											
New London, CT	33			28																		
New Haven, CT	35					17	16	20														
Bridgeport, CT	35					10																
Albany, NY	32		19	53																		
New York, NY	45															28	32	33	54	7	23	
Newark, NJ	40										43											
Camden, NJ	40(f)					10	20	6	4	2												

TABLE 3 (continued)

Port	Auth. Depth	30	31	32	33	34	35	36	37	38	39	40	41	42	43	44	45	46	47	48	49	50-60
Philadelphia, PA	40(f)										31	63	9	1	2	2	3	9	7	8	32	
Delaware R.	40(f)										124	122	37	16	15	13	9	20	16	12	64	
Wilmington, DE	40(f)					14	15	4	2	3												
Baltimore, MD	42											159	3									
Hampton Roads, VA	45															24	113	41	95			
Newport News, VA	45															7	73					
Norfolk, VA	45															17	50	41	87			
Wilmington, NC	38								51	7												
Morehead City, NC	40					64	39	6	6	1	1											
Charleston, SC	35																					
Brunswick, GA	30		25	1																		
Savannah, GA	38								47	13												
Jacksonville, FL	38								33	2												
Canaveral, FL	36					12	1															
Palm Beach, FL	33		30	5	5				3			2										
Port Everglades, FL	42																					
Miami, FL	38	272																				
Ponce, PR	30	54																				
St. Thomas, VI	33		2	2	1	98																
Tampa, FL	36						39															
Panama City, FL	32			1	2																	
Pensacola, FL	33			10	1																	
Mobile AL	40											157										
Pascagoula, MS	38							13	8													
Gulfport, MS	30	7																				
New Orleans, LA	40									381	415											
Baton Rouge, LA	40										111											
Beaumont, TX	40									114	31											
Port Arthur, TX	40									179	46	1										
Galveston, TX	40									227	73	3										
Texas City, TX	40									51	8											
Houston, TX	40									150	60											
Freeport, TX	38/36							100	168	11												
Corpus Christi, TX	45																					
Channel	47/45															4		1	1			
Harbor IS	47/45															3		1	1			
Corpus CH	40/45										35	29	25									

[a] No federal project
[b] Columbia River Bar limitation of 37 ft
[c] Maintained at 32 ft
[d] Port maintains 51 ft outer channels, 45 ft inner channels
[e] Port maintains 62 ft entrance channel, 50 ft and 55 ft inner channels
[f] Maintained at 35 ft

SOURCE: Data from U.S. Army Corps of Engineers, 1981.

TABLE 4 Port Calls of Bulk Carriers and Tankers[a] at Four Ports in 1980, by Actual and Design Drafts

| | (ft) | | Design Draft | | | Actual Draft | | | | Total |
Port	Existing Depth	Planned Depth	Ex. Dep. to 50	51 to 55	= or 56	or = 40	or = 42	41 to 50	46 to 50	Port Calls
Baltimore	42	50	165	12	29		897			974
Norfolk	45	55	126	81	53				141	1299
New Orleans	40	55	126	81	53			40		2621
Mobile	40	55	171	2	2	519				532

[a]Bulk carriers, oil/ore carriers, ore carriers, ore/bulk/oil carriers, and crude oil tankers.
Source: U.S. Maritime Administration, Office of Port and Intermodal Development.

TABLE 5 Approximate[a] Cargo Tons/Foot of Draft for Selected Vessels

Vessel: DWT (length x beam x draft, in ft)	Cargo (Long) Tons/Draft Ft
Panamax Bulk Carrier: 80,000 (850 x 106 x 49, fully loaded, or 40 for Panama Canal transit)	2300
Panamax Containership: 915 x 106 x 35	2260
Panamax Tanker: 80,000 (764 x 106 x 40)	2700
Bulk Carrier: 150,000 (915 x 145 x 55)	3380
Tanker: 390,000 DWT (1143 x 228 x 74)	6627
Bulk Carrier: 225,000 (1085 x 178 x 55)	4900

[a]Several factors affect actual cargo tons/ft of draft

TABLE 6 Port Calls by General Cargo Vessels by Draft, 1980 Foreign Trade

Vessel Type	Total Port Calls	Port Calls by Design Draft				Port Calls by Actual Draft				
		1-15	16-45	45-50	51+	1-15	16-45	45-50	51+	Other
Container-ship	8336	254	8082			333	8002			1
Freighter	14569	373	14195	1		1197	13364	6	1	1
Refrigerated Freighter	2352	75	2277			188	2163		1	
Car carrier	1180		1162		18	5	619			
Container/ car	19		19				19			
Container/ Ro-Ro	637	86	551			58	579			
Pallet carrier	43		43				43			
Partial container	6557	37	6520			199	6357		1	
Ro/Ro	3096	314	2782			611	2483	2		
Bulk/con- tainership	321		321			1	320			
Total	37110	1139	35952	1	18	2592	33949	8	3	2

SOURCE: U.S. Maritime Administration, Office of Port and Intermodal Development.

TABLE 7 Dry Bulk Carriers in the World Fleet, 1984 (including combination carriers), by DWT and Draft

Draft

DWT	0-29.9	30	31	32	33	34	35	36	37	38	39	40	41	42	43	44	45	46	47	48	49	50+
0-39,999	601	298	329	438	372	428	451	327	127	62	30	1										1
40,000					1	9	26	19	48	67	66	41	14	3	2							
50,000									9	18	33	133	56	1	1	1			1			
60,000								1		2	3	93	32	18	18	40	6	33	1			
70,000												5	4	114	49	29	41	18	37	7		
80,000											1		1	10	15	1	9		3	8	6	
90,000													1	1	7	1		3	10	8	7	7
100,000											1							1	3	7	24	26
110,000																					1	75
120,000																				1		77
130,000																						69
140,000								2														24
150,000																						38
160,000																						37
170,000																						5
180,000																						3
190,000																						6
200,000																						2
210,000																						5
220,000																						14
230,000																						1
240,000																						4
250,000																						
260,000																						7
270,000																						4
280,000																						
290,000																						
300,000																						
310,000																						
320,000																						
330,000																						
340,000																						
350,000																						
360,000																						
370,000																						
380,000																						
390,000																						
400,000																						
Total	601	298	329	438	373	437	477	349	184	149	134	273	108	147	82	72	56	55	55	31	38	404

SOURCE: MARDATA, Inc.

5090 bulk carriers
993 41 ft+   20%
404 50 ft+   8%

TABLE 7b Dry Bulk and Combination Carriers on Order, 1984, 1984, by DWT and Draft

DWT	Draft 0-29.99	30	31	32	33	34	35	36	37	38	39	40	41	42	43	44	45	46	47	48	49	50+
0- 39,999	31	14	75	63	44	111	150	47	18	44	12	11										
40,000-49,999						2	20	14	33	4		10	6	2	1							
50,000-59,999									2			24	5	24	12	2						
60,000-69,999								1								5	10					
70,000-79,999													1	1								
80,000-89,999													1									
90,000-99,999																		2				
100,000-100,999																						1
120,000-129,999																					2	3
130,000-139,999																						4
140,000-149,999																					2	6
150,000-159,999																						10
160,000-169,999																						3
170,000-179,999																						2
180,000-189,999																						1
240,000-249,999																						1
280,000-289,999																						1
300,000-390,999																						2
Total	31	14	75	63	44	113	170	62	53	48	12	45	13	27	13	7	10	2	0	0	4	33

SOURCE: MARDATA, Inc.

TABLE 8 Crude Oil Tankers in the World Fleet, 1984, by DWT and Draft

DWT	Draft 29.99	30-30.99	31	32	33	34	35	36	37	38	39	40	41	42	43	44	45	46	47	48	49	50+
< & inc. 39,999	1994																					
40,000-49,999		152	204	114	95	109	136	245	91	50	3	11	1	1	1	1			2	2		
50,000						1	7	10	15	43	30	7	6	11	5	5		1				1
60,000							1	5	5	14	40	51	19	30	29	10	2	1		1		1
70,000								1		5	4	17	28	12	24	9	5	1			1	1
80,000												21	33	11	27	32	25	1	2	2	1	
90,000												14	10	3		6	12	37	9	20	11	19
100,000												3	1			1	3	7	8	11	6	10
110,000																		5	4	3	12	18
120,000																					2	62
130,000																			2		1	82
140,000																						30
150,000																					2	26
160,000																						7
170,000																						20
180,000																						7
190,000																						1
200,000																						7
210,000																						30
220,000																						75
230,000																						55
240,000																						17
250,000																						75
260,000																						90
270,000																						65
280,000																						30
290,000																						5
300,000																						5

TABLE 8 (continued)

	1994																				Total
310,000																					27
320,000																					5
330,000																					8
340,000																					2
350,000																					16
360,000																					3
370,000																					1
380,000																					11
390,000																					3
400,000																					10
410,000																					5
420,000																					1
430,000																					-
440,000																					-
450,000																					2
460,000																					-
470,000																					3
480,000																					1
490,000																					-
500,000																					2
510,000																					-
520,000																					-
530,000																					-
540,000																					4
550,000																					1
Total	152	204	114	95	110	144	261	111	112	77	124	98	68	64	47	47	53	17	39	36	844

4821 total tankers
1323 - 41+ draft (27%)
1994 - <30' draft (41%)

SOURCE: MARDATA, Inc.

TABLE 8b Tankers on Order, 1984, by DWT and Draft

DWT	Draft 0-29.99	30	31	32	33	34	35	36	37	38	39	40	41	42	43	44	45	46	47	48	49	50+
0-39,999	135	3	8	8	7	13	14	16	2	3	3	2	2									
40,000-49,999	4						2	1			1				4							
50,000-59,999										2		7	1									
60,000-69,999										3		8		3	2	11						
70,000-79,999										2		10	2									
80,000-89,999										3		6		3		3			3			
90,000-99,999																	2			1		1
120,000-129,999																						1
130,000-139,999																1						
160,000-169,999																						1
230,000-239,999																						1
Total	139	3	8	8	7	13	16	17	2	13	4	33	5	6	6	15	2	0	3	1	0	4

TABLE 9 Containerships and Roll-On/Roll-Off Vessels in the World Fleet, 1984, by Draft and Length Over-All

Length Over-All	Draft 0-29.9	30	31	32	33	34	35	36	37	38	39	40	41	42	43	44	45	46	47	48	49	50+
0-619.9	12042	943	750	382	153	161	40	14	17	11	11	3	1	1	1						2	2
620-634.9	36	8	8	3	8	1	2	17	6	1	5	1			1							
635-649.9	19	1	8	3	6		10	8	1	2	5				1							
650-664.9	23	17	13	6	1	10	3		3	7												
665-679.9			12	9	12	2	3		3	1	1											
680-694.9	6	11	2	9	3	13	7	17	16	4	4		4	1								
695-709.9	13	7	1	10	3	7	4	4	2	2	1	1										
710-719.9						1	11	3	6	2												
720			1			11		1	6	2												
730	1	6	2			1			5	3	1	2										
740			3	12		1	14	1	2	1			3									
750					2	3		2		2				1								
760	1		4																			
770																						
780	4			1			2		5	1		1										
790									4				1		1							
800				1				2														
805											1											
810					4																	
815				1			4	2	1		5	1										
820																						
825		1				2	3															
830																1						
835	1																					
840																					1	1
845													1		5	5						
850									1		1		1			5						1
855										5	3											
860							3	2	2													
865																						
870																						
875																2						
880									1		1											
885																1						
890																						
895																						
900		2							1	3												
905																						
910																						
915																						
920																						
925																						
930																						
935																						
940															3	2						
945	1						8				2				2							
950																8						
Totals	12147	996	804	437	192	213	111	74	81	45	40	9	10	3	14	19					3	2

15,200 Ro-Ros and ships that carry containers
   411 35 ft+ (3%)

SOURCE: MARDATA, Inc.

TABLE 9b Containerships and Roll-On/Roll-Off Vessels on Order (1984)

LOA-	0-29.9	30	31	32	33	34	35	36	37	38	39	40	41	42	43
0-619.9	90		2	4	1	5		2		2					
620-634.9	2														
635	2								3						
650	6	2							1		4				
665	5					4					1				
680											1				
695															
710															
720															
730															
740															
750						2									
760										9					
770															
780															
790															
800															
805															
810															
815			1				2								
820							1								
825															
830															
835															1
.															
.															
855-859.9										3					
.															
.															
885-889.9														2	
.															
.															
945-949.9	12														
Totals	117	2	3	4	1	11	3	2	4	14	6			2	1

170 containerships and Ro/Ros on order
32 - 36 ft draft or more (19%)
SOURCE: MARDATA, Inc.

TABLE 10 Freight Rates for Bulk Carriers

Highest and Lowest Rates*		in $/ton of cargo		
		1981	1982	1983
Grain	High	22.00	12.00	9.00
U.S. Gulf ports to	Low	8.75	5.74	7.00
Rotterdam or Antwerp				
Coal				
Hampton Roads	High	28.50	19.60	17.50
to Japanese ports	Low	17.50	10.80	12.35
Iron Ore				
Brazil to Northwest	High	15.00	7.00	6.50
European ports	Low	7.00	4.45	5.95

SOURCE: Maritime Transport Committee, 1984.

Transportation Savings: Cost vs. Price*

Coal		$/ton Panamax		125,000 DWT	
		1980	1983	1980	1983
U.S. East Coast to Rotterdam/Antwerp	Cost	12.58	10.97	9.64	8.57
	Savings for Larger Vessel			2.94	2.40
	Market Rate	10.29	5.51	7.45	4.12
	Savings for Larger Vessel			2.84	1.39
U.S. East Coast or Gulf to Japan	Cost	40.23	35.993	28.63	25.35
	Savings for Larger Vessel			11.60	10.58
	Market Rate	33.08	18.08	21.91	11.57
	Savings for Larger Vessel			11.17	6.51

SOURCE: Poten & Partners, 1983.

TABLE 11 World Ports Capable of Handling 150,000 DWT (+) Vessels

North Pacific	North Atlantic
Nigata	Narvik
Mizushima	Foulness
Kurf	Heligoland
Kashima	Clyde Port
Kimitsu	Glasgow
Chiba	Tees-Port
Oita	Bantry Bay
Kiire	Liverpool
Tsurusaki	Milford Haven
Okinawa	Bilbao
Tokyo Bay	Gijon
Kawasaki	Algeciras
Yokkaichi	Gothenburg
	Port Talbot
South Pacific	Hamburg
Port Hedlund	Dunkirk
Dampier	Rotterdam/Europoort
Hay Point	Le Havre
Caves Beach	Zeebrugge
Sydney	Antwerp
Clutha	
Kembla	Mediterranean
Bonython	Fos/Marseilles
	Genoa
North America (exec. USA)	Marsa El Breg
Roberts Bank	Taranto
Seven Islands	Trieste
Come-by-Chance	Port Said
Point Tupper	
St. John	Persian Gulf
	Ras al Khafji
South America	Ras Tanura
Bolivar	Mina al Ahmadi
Puerto La Cruz	Kharg Island
Sepetiba	Das Island
Tubarao	
Huasco	South Africa
San Nicolas	Richards Bay
	Port Elizabeth
	Algoa Bay
	Saldanha Bay

TABLE 12 Port Improvement Activities Worldwide (Responses to Query of Committee on National Dredging Issues)

Country	Port	Entrance Channel Width m	(ft)	Depth m	(ft)	Expansion
Australia	Kembla	305	(1000)	16.75	(54.0)	Physically not practical
	Melbourne	125	(410)	13.1	(43.0)	-
	Newcastle					Expansion possible, but not planned to accommodate tankers up to 200,000 DWT
	Bonython			20.0	(65.6)	Not practical
Belgium	Antwerp	500	(1640)	17.75	(58.2)	Underway
	Zeebrugge	300	(984)	18.00	(59.0)	Underway
Brazil	Santos	150-250	(492-820)	14.0	(45.9)	Expansion possible
	Tubarao	280	(191)	22.5	(73.8)	No plans at present
Canada		200-250	(656-820)	10.0	(33)	-
	Montreal	244	(800)	10.7	(35.0)	-
	St. John N.E.			10.4	(34)	No plans at present
	Vancouver	305	(1000)	15.2	(50)	
	Roberts Bank	457	(1500)	19.5	(64)	Deep port recently completed
	Halifax			22.9	(75)	No plans
China	Su-Ao	250	(820)	22	(72.2)	Need exists for a new deep water port
Denmark	Aalborg	280	(919)	-		Extension planned 1984-86
	Arhus	-		14	(45.9)	Recently completed (1983)
Ecuador	Guayaquil	122	(400)	9.5	(31)	Planned
England	Tyne					Coal terminal planned
	Southampton	335	(1,100)	12.7	(41.5)	No plans at present
	Port Talbot	168	(550)	9.5	(31.0)	No plans at present
	Immingham	213	(700)	8.8	(29.0)	No plans at present
	Southampton	325	(1066)	12.8	(42.0)	No immediate plans
Finland	Helsinki	250-350	(820-1148)	9.6	(31.5)	No plans at present
France	Rouen	100-200	(328-650)	12.0	(39.4)	No
				Maximum		
Germany	Hamburg	250	(820)	13.5 16.5 (at high tide)	(44.0) (54.0)	Plan new deep water ports
India	Bombay	Naturally wide access channel		10.0 13.3 (at high tide)	(32.8) (43.6)	Construction of a new deep-water port underway
	Mormugao	250	(820)	13.7 18-20 (outer harbor)	(45.0) (59-65.6)	Proposal to deepen to 16.5 m x (54 ft) under consideration
	Kandla	183	(600)	10.4	(34.1)	-
	Kandla Offshore Oil Terminal	Naturally wide access channel		31.0 37.0 (at high tide)	(101.7) (121.4)	Construction underway
Indonesia	Semarang	60	(197)	5	(16.4)	Deepening to 9 m (29.5 ft) planned
Ireland	Cork					Development of a deep-water harbor stopped (1983)

TABLE 12 (continued)

Country	Port	Entrance Channel				Expansion
		Width		Depth		
		m	(ft)	m	(ft)	
Italy	Voltri					Development of a deep-water port delayed
Jamaica	Kingston	244	(800)	11.1	(36.5)	-
Japan	Nagoya			12.0	(39.4)	Deepening completed (1984)
Kenya	Mombasa	300	(984)	20.0 36.0 (depth in shelter area)	(65.6) (118)	Plan to construct a deep-water port to Lamu
Kuwait	Shuwaikh					Major expansion underway, to be completed 1986
	Malaga					Expansion planned for the next 10 years
Malaysia	Tanjung Berhala					Major port handling ships up to 220,000 DWT to be completed in 1984
Netherlands	Rotterdam-Europort	600	(1968)	27.5	(90.0)	Deepening to allow ships of 350,000 DWT to enter port
New Zealand	Auckland	300	(984)	11.0	(36.1)	No evidence for the need of a deeper port
	Taranaki					Being deepened, completion (1986)
	Tauranga	2500	(8202)	11.3 (at low tide)	(37.0)	No
Nicaragua	New Port			12.0	(39.4)	Deep water port under construction
Saudi Arabia	Yanbri	200	(656)	12.0	(39.4)	Five general ports under construction. Completion 1985
Spain	Santander			13.0	(42.7)	Under construction
South Africa	Richards Bay	400	(1312)	19.5	(64.0)	Deepening to 23 m depth underway
	Saldanha Bay	500	*1640)	23.0	(75.5)	
	Durban	183	(600)	12.7	(41.7)	
	Port Elizabeth	310	(1017)	12.2	(40.0)	
Sri Lanka	Colombo	230	(754)	11.5	(37.7)	No plans
Taiwan	Kaohsiung	150	(492)	16.0	(52.5)	No plan at present
	Keelung	276	(906)	20.0	(65.6)	
	Taichung	300	(984)	20.0	(65.6)	Deeper port planned
Thailand	Mabtapud					To be constructed in future
	Laem Chalang					Construction to start in 1987
UAE, Dubai	Jebel Ali	280-235	(919-771)	15.0	(49.2)	Expansion plans curtailed
	Ras Al Khaimah			14.2	(46.6)	

TABLE 13 Estimated Costs and Trade by Selected Ports, 1990

Port	(million 1990 dollars)			(million short tons)		
	Existing Operations and Maintenance Costs[c]	Total Costs, New Construction Dredging[d]	Incremental Operations and Maintenance Costs[e]	Estimated Total Trade[f] in 1990	Estimated Coal Exports in 1990	Deepening Plans
Hampton Roads[a]	3.2	438.5	6.1	80.6	58.9	55 ft
Great Lakes[b]	4.5	0.0	0.0	31.5	19.7	None
Baltimore	2.1	383.7	1.6	74.9	29.2	50 ft
New Orleans/ Baton Rouge	14.9	479.6	125.1	173.5	8.6	55 ft
Mobile	4.6	371.8	2.8	25.8	4.7	55 ft
Los Angeles/ Long Beach	0.1	420.2	0.0	81.4	2.8	80 ft
Philadelphia	5.8	0.0	0.0	67.7	12.0	None

[a]Norfolk and Newport News, Virginia
[b]Includes Ohio ports of Ashtabula, Conneaut, Sandusky, and Toledo
[c]Converted from 1982 dollars using GNP deflator 1.641
[d]Converted from 1981 dollars using GNP deflator 1.0946
[e]Converted from 1981 dollars using GNP deflator 1.0946
[f]Exports, imports, and coastwise movements

SOURCES: Office of Policy, U.S. Army Corps of Engineers, for cost estimates. 1990 total trade by port estimated from U.S. Maritime Administration data and forecasts from the Federation of American Controlled Shipping. 1990 coal trade by port estimated using regional coal export forecasts from the International Coal Trade Model, existing port capacity and capacity under construction from U.S. Maritime Administration.

TABLE 14 Planning, Approval, Authorization, and Funding Process for Major Navigation Projects.

---

1. Congress authorizes study.
2. Congress appropriates funds.
3. Following appropriation of funds, District Engineer conducts initial public meeting to review draft plan of study. This provides opportunity to identify and discuss local problems and alternatives emphasizing national economic efficiency and environmental quality.
4. District Engineer
   - Investigates all alternatives
   - Performs limited
     - technical feasibility studies
     - environmental assessments
   - Proposes most feasible solutions in preliminary feasibility report.
5. Formulation stage--Stage public meeting to discuss most feasible alternatives.
6. District Engineer
   - Investigates formulation stage alternatives
   - Performs detailed
     - technical feasibility studies
     - environmental assessments
   - Selects plan for proposal in detailed Feasibility Report (FR)
   - Distributes draft Environmental Impact Statement (EIS) & FR (15 days prior to state public meeting)
   - Files draft EIS with EPA.
7. Public meeting--tentative plan proposed and discussed.
8. States, agencies, interest groups, public respond to draft EIS and draft FR.
9. District Engineer
   - Reviews comments to draft EIS & FR
   - Prepares recommended
     - Final EIS
     - Final FR.
10. Division Engineer
   - Reviews
   - Modifies as Appropriate
     - Final FR as Appropriate
     - Final EIS
   - Issues public notice requesting public views be sent to Board of Engineers for Rivers & Harbors (BERH)
   - Forwards recommendations to BERH.
11. Board of Engineers for Rivers & Harbors
   - Considers Views of
     - Public
     - States
     - Agencies

TABLE 14   (continued)

_____

    o Reviews and provides recommendations
      - Final EIS
      - Final FR.
    o Transmits to chief of engineers.
12. Chief
    o Reviews Board report
    o Prepares his draft recommendations
    o Distributes for outside review
    o Files final EIS with EPA
    o Circulates to public for 30-day review period and to governors,
      federal departments (90-day review period).
13. Chief
    o Reviews comments received
    o Modifies report as appropriate
    o Prepares record of decision (ROD).
14. Chief
    o Forwards recommendations to Secretary of the Army for
      consideration
      - Final Report
      - Final EIS
      - ROD.
15. Secretary of the Army
    o Reviews
    o Coordinates with OMB
    o Prepares his recommendations
    o Forwards final FR, final EIS
    o ROD to Congress (6 mo.).
16. Project Authorization
    o Congress holds hearings
    o Congress includes in Water Resources Development Act or other
      legislation
    o President signs.
17. OMB
    o Reviews Corps budget
    o Submits to Congress.
18. Project Funding
    o Congress includes in Appropriations Act
    o President Signs.
19. Local interests guarantee to fulfill obligations required by law
    (e.g., real estate, cost sharing, maintenance, operation, flood
    zoning).
20. District Engineer
    o Formulates pre-construction planning general design memoranda
      - Updates EIS as required for Sec. 404 compliance, obtains
        necessary Water Quality certificates
      - Issues public notice and conducts at least one public meeting
        (36 mo.)
    o Obtains additional congressional authorization as appropriate
      (24 mo.)
    o Initiates and completes construction (60 mo.)
    o Operates and maintains.

TABLE 15 Average schedule for navigation projects, U.S. Army Corps of Engineers

ACTIVITY				
Survey/study authorized				
Funds for study appro- priated	4.9 years			
Study/survey sent to divi- sion		4.1 years		
Report sent to Congress			1.5 years	
Project authorized			0.6 year	
Initial funds appropriated for preconstruction planning & engineer- ing			1.9 years	
Initial construction funds appropriated				5.8 years
First contract award				2.8 years
YEARS				TOTAL TIME 21.6

SOURCE: General Accounting Office, 1984.

TABLE 16 Year of Authorization of Main Channels of Selected Ports

State	Port and Project	Date	Channel Size and Type
Alaska	Anchorage	July 3, 1958	35'x irregular berthing area
		Oct. 22, 1976	35' berthing area relocated
Alabama	Mobile	Sept. 3, 1954	42'x 600' entrance, 40'x 400' bay
		July 26/ Dec. 15, 1970	40'x 400' extension to Theodore
California	Humboldt Harbor and Bay	July 16, 1952	40'x 500' entrance
		Aug. 13, 1968	35'x 400' bay
	Stockton (J.F. Baldwin overlays this and other projects)		
	San Joaquin R.	Aug. 26, 1935	30'x 400' lower/225' upper land cut and river
	J. F. Baldwin	Oct. 27, 1965	35'x 400/225' (under construction)
	Suisun Bay Channel		
		Jan. 21, 1927	30'x 300' bay
	J. F. Baldwin	Oct. 27, 1965	45' lower bay, 35' upper bay (under way) (35'x 400' lower and upper, per redesign)
	San Pablo Bay	Jan. 21, 1927	30'x 700', Mare Island Strait 30'x 400' Pinole Shoal
	J. F. Baldwin	Oct. 27, 1965	45' Pinole Shoal (35'x 400' per redesign)
	Oakland	March 2, 1945	35'x 800/950' outer basin
		Oct, 23, 1962	35'x 600' inner harbor
	Richmond	Sept. 3, 1954	35'x 600' bay approach and inner harbor
	J. F. Baldwin	Oct. 27, 1965	45' maneuver basin at Long Wharf (construction/design under way)
	San Francisco (most piers on deep water, except Islais Creek)		
		Aug. 30, 1935	35' bay shoal (Islais approach)
	J. F. Baldwin	Oct. 27, 1965	55'x 2000' bar and entrance (completed) 45' bay shoals (small, completed) 45'x 600' Southampton Shoal design under way, present size 25' x 600')
	Los Angeles	Oct. 22, 1976	45'x 1000' entrance, 45'x 750 outer 45' basins
	Long Beach	July 3, 1930	35'x 300/500' entrance, 35'x 400/ 1200' basins
	(LA-LB Project)		(Local overdredging to 52' LA, 65' LB)
	San Diego	July 3, 1930	40'x 800' entrance, 35'x 1500/5200' inner basins
		Aug. 30, 1935 (widening)	30'x 2200', bay (Navy overdredging to 42')

TABLE 16 (continued)

State	Port and Project	Date		Channel Size and Type
Connecticut	Bridgeport	July 3,	1950	35'x 400' entrance and main harbor
	New Haven	July 24,	1946	35'x 500' entrance, 35'x 800/400 harbor
	New London	Aug. 26,	1937	33'x 600' entrance and harbor
		Oct. 22,	1976	40' design only, restudied, deferred
Delaware	C&D (Canal (Inland water-way: Delaware River, Chesa-peake Bay)	Sept. 3,	1954	35'x 450' land cut and Chesapeake Bay (Latter connects with Baltimore 35'x 600')
	Wilmington	July 14,	1960	35'x 400' at Delaware R. upper reaches shallow
	New Castle & Delaware City (Delaware R.-Phila-delphia to sea)	June 20,	1938	40'x 1000' entrance, 40'x 800' to Philadelphia
Florida	Charlotte	May 17,	1950	32'x 300' bar and entrance (to Boca Grande)
		July 3,	1930	10'x 100' inner (Boca Grande/ Punta Gorda)
	Canaveral	Oct. 23,	1962	37'x 400' entrance, 36x 300' inner, 36' basins
	Panama City	June 30, (June 14,	1948 1972)	34'x 450' entrance, 32'x 300' bay [42'x 450' entrance, 40'x 300' bay, 38' inner and basins (not built)]
	Port St. Joe	Sept. 3,	1954	47'x 500/400' entrance, 35'x 300' bay
	Pensacola	Aug. 27,	1962	35'x 500' entrance, 33'x 300' bay, 33'x 500' inner
	Palm Beach	July 14,	1960	35'x 400' entrance, 33'x 300' bay 33' basins
	Jacksonville	Oct. 27,	1965	42/40'x 400' entrance, 38'x 400/ 1200' lower river
		March 2,	1945	34'x 400/1200' upper river (above Blount Island
	Key West Harbor	July 25,	1912	30'x 300' entrance, 26'x 800' basin
	Tampa Harbor	Dec. 31,	1970	46'x 700' bar, 44'x 600' entrance, 44'x 500' bay, 42/40'x 400/300' inner
	Miami Harbor	Aug. 13,	1968	38'x 500' entrance, 36'x 400' bay, 36' basins
	Port Everglades	May 9/31,	1974	45'x 500' entrance, 42'x 450' bay 42' basins
Georgia	Brunswick	May 17,	1950	32'x 500' bar, 30'x 400' entrance and upstream

TABLE 16 (continued)

State	Port and Project	Date	Channel Size and Type
	Savannah	Oct. 27, 1965	40'x 600' bar, 38'x 500' entrance and lower river, 38/36'x 400 upper river
Hawaii	Port Allen	Aug. 30, 1935	35'x 500' entrance, 35' basin
	Nawiliwili	Sept. 3, 1954	40'x 600' entrance
		March 2, 1919	35' basin
	Kahului	June 25, 1910	35'x 600' entrance, 35' basin
		July 14, 1960	35' basin extension
	Hilo	March 3, 1925	35'x 1400' basin, breakwater protected
	Honolulu	Oct. 27, 1965	45'x 500' entrance, 40' basins
		Sept. 3, 1954	35'x 400' second entrance
Louisiana	Lake Charles	July 14, 1960	42'x 800' bar, 42/40'x 400' entrance
	(Calcasieu R. & Pass		40'x 400' river and cut, 35 x 250 above Lake Charles
	New Orleans	March 2, 1945	40'x 600' SW Pass Bar, 40'x 800' SW Pass
	Baton Rouge		30'x 600' SW Pass Bar, 30'x 450' SW pass
	(Miss. R.-Baton		40'x 1000' between New Orleans and Passes
	Rouge to Gulf)		(35' at and above New Orleans)
		Oct. 23, 1962	40'x 500' New Orleans to Baton Rouge (overlays prior 35 x 1500 at New Orleans)
Maine	Portland	Oct. 23, 1962	45'x 1000' entrance, 45' anchorage
		Aug. 8, 1917	35'x 1000/400 inner channels
Maryland	Baltimore	July 3, 1958	42'x 1000/800' bay entrance & shoals
			42'x 800' harbor entrance & main channel
			42'x 600', 42/35 x 400, 39 x 400 side channel
			35'x 600' bay channel to C&D canal
		(Dec. 31,1970)	[50'x 1000' bay entrance & shoals]
			50'x 800' harbor entrance & main channel
			50'x 600', 42/35'x 400', 40/49'x 400 side channel
			(deepening not started on 1970 project)]
Massachu-setts	Cape Cod Canal	Jan. 21, 1927	32'x 540/480' land cut, 32'x 700/500' bay approach (south end)
	Fall River	July 24, 1946	35'x 400' bay and river
		Sept. 3, 1954	35'x 400' bay side channel
		(Aug. 13,1968)	[40'x 400' all of the above (not built)]

TABLE 16 (continued)

State	Port and Project	Date	Channel Size and Type
	Boston Harbor	Aug. 30, 1935	40'x 900' (45' in rock) + 35'x 600' adjoining, main entrance; 30'x 1200' auxiliary entrance 40'x 600' + 35'x 600' adjoining inner channels 35' connections w/Chelsea, Charles, Mystic R.
		July 13, 1892	27'x 1000' auxiliary entrance, partly overlays 35'x 500' Weymouth Fore R. entrance
Massachu-setts	Boston (other projects entered via Boston Harbor project)		
	Mystic River	May 17, 1950	35'x 1000' $\pm$
	Chelsea River	Oct. 23, 1962	35'x irregular
	Dorchester Bay	Oct. 23, 1962	35'x 300'
Mississippi	Gulfport	June 30, 1948	32'x 300' bar, 30 x 220 bay
	Pascagoula	Oct. 23, 1962	40'x 350' bar, 38'x 350' bay, 38'x 225' bay to Bayou on Casotte (refinery)
New Hampshire	Portsmouth (Portsmouth Harbor & Pis-cataqua River)	Sept. 3, 1954	35'x 400' river/strait channel w/ widened bends (natural entrance)
New Jersey	Camden (Delaware R. at Camden)	March 2, 1919	30'/18'x 800' $\pm$ (adjoins 37' portion of Delaware R. Phila-delphia to sea project)
		(Mar. 2, 1945)	[37'x 800' access to marine terminal (not built)]
	(Del. R. Phila. to Sea)	June 20, 1930	40'x 1000' entrance, 40'x 800' to Philadelphia-Camden, 37'x 1000' at Philadelphia-Camden Camden, 37'x 1000' at Phila-delphia-Camden (west half of channel now 40'- see Phila-delphia)
	Gloucester (see Camden projects)		30' portion of Camden project
	Paulsboro (Delaware R.-Philadelphia to sea)	June 20, 1938	Direct access to/from 40'x 800' channel
New York	Albany (Hudson R., NY)	Sept. 3, 1954	32'x 600' lower river (to Kingston), 32'x 400' upper (34' in rock)

TABLE 16   (continued)

State	Port and Project	Date	Channel Size and Type
New York/ New Jersey	New York (Port includes 16 projects deeper than 14')		
	NY Entrance & Anchorages	Aug. 26, 1937	45'x 2000' entrance (Ambrose channel)
			45'x 2000' anchorages (Upper Bay)
		Aug. 30, 1935	35'x 800' entrance (Sandy Hook)
		July 5, 1884	30'x 1000' Sandy Hook/Ambrose Channel connection
	Hudson R., NY & NJ	Aug. 26, 1937	45'x 2000' w/ 48'x 2000' at upper end, 40'x 30' side channels (Weehawken, etc.)
	NY & NJ Channel	May 28, 1935	35'x 600/500' Sandy Hook Bay & Arthur Kill 35'x 800/1000' in Kill van Kull (37' in rock)
		Oct. 27, 1965	Kill van Kill widened
	East River	Sept. 22, 1922	40'x 1000', 35'x 550' upper, 35'x 1000' at Long Island Sound
	Buttermilk Channel	Jan. 13, 1902	35'x 500' Adjacent Brooklyn
		May 1935	40'x 500' channels Waterfront &
	Bay Ridge & Red Hook channel	March 3, 1899	40'x 1200' & East R./
		July 3, 1930	40'x 1750' Upper Bay Connections
	Newton Creek	July 3, 1930	23'x 130' East R./Brooklyn
	Wallabout Chan	March 3, 1899	20'x 230/350' side channels
	Gowanus Creek	July 16, 1952	30'x 500/200'
	Newark Bay	Mar. 22, 1945	35'x 700' bay (500' above Port Newark) 37' at turn connecting w/Kill van Kull (access to sea via NY & NJ channel project)
	Hackensack R.	Sept. 3, 1954	32'x 400/300' (34' in rock)
	Passaic R.	July 3, 1930	30'x 300' lower river
North Carolina	Morehead City	Dec. 31, 1970	42'x 450' entrance, 40x 400/600 bay
	Wilmington	May 17, 1950	40'x 500' bar & entrance 38'x 400' river
Oregon	Yaquina Bay and Harbor	July 3, 1950	40'x 400' entrance, 30'x 300' bay
	Coos Bay	Dec. 30, 1970	45'x 700' entrance, tapers to 35'x 300 35'x 300 bay (w/bend widenings)
	Portland (Columbia R. Mouth)	Sept. 3, 1954	48'x 1/2 mile bar channel
	(Columbia R. & Lower Willamette)	Oct. 23, 1962	40'x 600' river
	Astoria (same as Portland, plus 40'x 800' basin)		

TABLE 16 (continued)

State	Port and Project	Date	Channel Size and Type
Pennsyl-vania	Penn Manor (Del. R. Phila-delphia /Trenton)	Sept. 3, 1954	40'x400' river Philadelphia to Fairless Works (35x 300, 12x 200 above Penn Manor)
	Delaware R. Philadelphia to Sea	June 20, 1938	40'x 1000' entrance, 40'x 800' to Philadelphia 40'x 400/500 west side + 37'x 500/600' east side (adjacent channels at Philadelphia-Camden
	(Schuylkill R.)	July 24, 1946	33'x 400', 33'x 300', lower river
		Aug. 8, 1917	26'x 200, 22'x 200', upper river
	Chester & Marcus Hook (Delaware R. Philadelphia at sea)	June 28, 1938	Direct access to/from 40'x 800' channel to Philadelphia (40'x 1000' bay entrance)
Puerto Rico	Mayaguez	Aug. 30, 1935	30'x 1000/500' entrance
	Ponce	Sept. 23, 1976	36'x 600' entrance,
		Oct. 1, 1976	36'x 400' inner
	San Juan	Aug. 4, 1976	48'x 800' bar, 46/40'x 800' bay, 40'x 400', 450' inner channels
South Carolina	Charleston	Oct. 17, 1940	35'x 1000' entrance, 35'x 600/400' Cooper River
		(July 18, 1918)	[40' conditionally authorized, not built]
		(Oct. 22, 1976)	[Phase I design for 42' entrance, 40' river]
	(Shipyard R.)	March 2, 1945	30'x 300/200'
	(Ashley R.)	Aug. 26, 1937	30'x 300'
Texas	Brownsville (Brazos Is)	May 17, 1950	38'/36'x 300' bar & entrance, 36'x 200' bay and land cut, 36 x 300/500 upper
	Matagorda	July 3, 1958	38'x 300' bar & entrance, 36'x 200' bay
	Freeport	May 17, 1950	38'x 300' bar, 36'x 200' entrance 36'x 200/400' inside, 30 x 200 side channel
		(Dec. 31,1970)	[47'x 400' bar, 45'x 400' entrance 45'x 400/375 inside, 30 x 200, 36 x 200 sides (1970 project not built)]

Galveston Harbor and Channel (channel is Gulf entrance for Galveston, Houston, Texas City)

		July 3, 1958	42'x 800' outer bar, 40'x 800' inner bar and bay entrance
		June 23, 1971	40'x 1125' harbor channel

204

TABLE 16   (continued)

State	Port and Project	Date	Channel Size and Type
	Houston	July 3, 1958	40'x 400' bay and lower 1/2 land cut
			40'x300' upper land cut to Sims Bayou
		June 30, 1948	36'x 300' above Sims Bayou (Manchester)
	Texas City	July 14, 1960	40'x 400' bay
	Corpus Christi	Aug. 13, 1968	47'x 700' outer bar, 45'x 600' jetty entrance, 45'x 600/500/400' bay
			45'x 300/400' to La Quinta (Reynolds)
			45'x 300/400' harbor & basins
	Sabine Pass	Oct. 23, 1962	42'x 800' bar, 40'x 800/500' entrance
	Port Arthur		40'x 500' pass & land cut (Port Arthur)
	Beaumont		40'x 400' Neches R. (Beaumont)
	Orange (Sabine-Neches Waterway)	Sept. 3, 1954	30'x 200' Sabine R. (Orange)
Virginia	Norfolk (Thimble Shoal's channel is entrance for Norfolk, Newport News and other ports) Norfolk project bay channel also used by Newport News and other ports)		
	(Thimble Shoals)	Oct. 27, 1965	45'x 1000' bay channel with 32 x 450 adjacent channels (both sides) part way
	(Norfolk)	Oct. 27, 1965	45'x 1500' bay, 45'x 800' to Lamberts Point
			40'x 750/450' above Lamberts point and up Elizabeth R., lower South Branch
		Oct. 22, 1976	35'x 250' Elizabeth R., upper South Branch
		March 2, 1907+	25'x 500/200 Elizabeth R., East Branch
	Newport News (Channnel to Newport News)	Oct. 27, 1965	45'x800'
Washington	Grays Harbor	March 2, 1945	30'x600' bar, 30'x350' bay
	Chehalis R.	Sept. 3, 1954	30'x200' river
	Kalama & Longview		
	(Col. R. Mouth)	Sept. 3, 1954	48'x1/2 mile bar channel
	(Col. R. & Lower Willamette)	Oct. 23, 1962	40'x600' Columbia R.

TABLE 16   (continued)

Port	Project	Date	Channel Size and Type
	Vancouver (on Columbia but with three different depths)		
	(Col. R. Mouth)		48'x1/2 mile bar channel
	Col. & Lwr Willamette)		40'x600' to mile 1055, 35'x500' above
	(Col. R. Vancouver to the Dalles)	Aug. 26, 1937	27'
	Everett (Everett Hbr & Snohomish R.)	July 3, 1930	30'x700/900' to bay waterfront
	Bellingham	July 3, 1930	26'x200' Squalicum Waterway
		July 3, 1958	30'x363.2' Whatcom Waterway
	Seattle	March 2, 1919	34'x750' West Waterway 34'x750/400 East Waterway (local overdredging to 40')
		March 3, 1925	30'x200' lower, 20x150, 15x150 upper Duwamish R.
	Port Angeles (Ediz Hook)	March 7, 1974	Breakwater project w/12'-15' small boat basin. 30' shoal removal deauthorized
	Tacoma	June 13, 1902	29'x500', 22'x500', 19'x500/250' City WW (deauthorization proposed)
		Sept. 3, 1954	35'x300' N1/2, 30'x350' S1/2 lower WW 35'x600/300 upper Blair WW
		Aug. 26, 1927 & July 3, 1930	30'x200' w/widenings, Hylebos waterway

Note: Authorization dates shown are the earliest dates with the specified channel(s) at present depth and width. Subsequent authorizations for extensions, bend widenings and other minor modifications not shown.

Dimensions for channels with tapering width or depth shown with a slash. Dual-depth channels have separate dimensions shown for each side (i.e., Boston, Philadelphia). A series of dimensions is shown for channels that taper in steps.

SOURCE: U.S. Army Corps of Engineers, Office of Civil Works.

TABLE 17 Proposals for Dredging to Depths Between 40' and 46'

Port	Existing Depth	Proposed Depth	Cost (10^6)	Status
Fall River, Mass.	35	40	$ 66	Approved; deferred
New London, Conn.	33	40	unav.	Recommended by federal study; being reviewed for approval
New Haven, Conn.	35	40	$ 23	"
Bridgeport, Conn.	35	40	unav.	Being studied
Newark, N.Y. (Kill van Kull)	35	40	$229	Recommended by federal study; being reviewed for approval
Howland Hook (Arthur Kill) N.Y.	35	40	$ 15/ $ 26	Being studied
Gowanus Creek Channel, N.Y.	30	40	$ 3	Recommended by federal study; being reviewed for approval
Port Jefferson, N.Y.	16	40	unav.	Approved; deferred
Elizabeth R., Norfolk, Va.	30/35	40/45	$ 5	Approved; under way
Elizabeth R., Norfolk, Va.	35/40	40/45		Recommended by federal study; being reviewed with other Norfolk deepening proposal (cost included in over-all proposal)
Charleston, S.C.	35	42	$ 73	Being studied
Savannah, Ga.	38	40	$ 30/ $ 80	"
Jacksonville, Fla.	38	44	$160	Being studied
Ft. Pierce, Fla.	25	40	$ 51	"
San Juan, P.R.	36	40	$ 65	Recommended by federal study; being reviewed for approval
Tampa, Fla.	34	43	$178	Approved; nearing completion
Charlotte Harbor, Fla.	32	40	unav.	Being Studied
Freeport, Tex.	36	45	$ 90	Approved
Corpus Christi, Tex.	40	45	$ 90	Approved; under way

TABLE 17 (continued)

Port	Existing Depth	Proposed Depth	Cost (10^6)	Status
Brownsville, Tex.	36	42	$ 23	Recommended by federal study; being reviewed (port has explored private financing)
Grays Harbor, Wash.	30	46	$ 71	Recommended by federal study; being reviewed
Everett	30	40	unav.	Being studied
Blair Waterway Tacoma, Wash.	35	45	$ 30	Recommended by federal study; being reviewed
Sitcum Waterway Tacoma, Wash.	35	40	$ 32	"
San Pablo Bay, Calif.	35	45	$166	Approved; under way
Oakland, Calif.	35	42	$ 38/ $?	Outer harbor ($38 million) recommended; inner harbor deepening being studied
Richmond Calif.	35	41	$ 51	Recommended by federal study; being reviewed for approval
Honolulu, Hawaii	35	40	unav.	Approved; underway
Hilo, Hawaii	35	40	$ 4	Being studied
Apra Harbor, Guam	35	40	$ 4	"
Baltimore, Md.	42	50	$420	Approved; no appropriation
Norfolk/ Newport News, Va.	45	55	$480	Recommended by federal study; being reviewed for approval
York R., Va.	22	50	$500	Being studied
Mobile, Ala.	40	55	$407	Recommended by federal study; being reviewed for approval
Pascagoula, Miss.	38	55	unav.	Being studied
New Orleans/ Baton Rouge, La.	40	55	$525	Recommended by federal study; being reviewed for approval
Sabine-Neches Waterway, Beaumont, Port Arthur, Orange, Sabine Pass Harbor, Tex.	40	50	$344	Being studied

TABLE 17 (continued)

Port	Existing Depth	Proposed Depth	Cost (10^6)	Status
Galveston, Tex.	40	55	$139	Being studied; port has sought private funding-- now completing new environmental impact statement required by court decision
Texas City, Tex.	40	50	$167	Recommended by federal study; being reviewed for approval
Houston, Tex.	40	50	$270	Being studied
Freeport, Tex.	36	50	unav.	"
Corpus Christi, Tex.	40	50	unav.	"
Columbia R. Bar, Ore., Astoria, Ore., Kalama, Longview, Vancouver, Wash.	48	60[a]	unav.	"
Astoria, Ore.	40	50	unav.	"
N.Y. Harbor & adjacent channels, Stapleton/ Port Jersey	45/35	70/60	$413	Being studied
Delaware R., transshipment facility, Pa.	40	90	unav.	"
Los Angeles/ Long Beach, Calif.	55/60	80	$460	Being studied
San Francisco, Calif.	45	55	unav.	Approved; under way

[a]Will not result in 60' port channel - bar subject to waves and swell; greater depths required - would be compatible with Columbia River depths of perhaps 50'.

SOURCE: Heiberg (1983).

TABLE 18   Cost of Maintenance Dredging

State/ Port	Annual Average Maintenance Cost ($ thousands)	Cost/ Cargo Ton	State/ Port	Annual Average Maintenance Cost ($ thousands)	Cost/ Cargo Ton
Alaska			Mississippi		
Anchorage	1453.6	.83	Gulfport	1899.2	1.53
Alabama			Pasacagoula	2485.5	.10
Mobile	5303.2	.83	New Hampshire		
California			Portsmouth	140.7	.05
Humboldt	1243.9	.94	New Jersey		
Long Beach	72.0	.00	Camden	264.4	.09
Los Angeles	72.0	.00	Gloucester	245.4	
Oakland	1143.3	.16	Paulsboro	2522.9	
Redwood City			Trenton		
Richmond	118.21	.06	New York/		
Sacramento			New Jersey	12905.7	.13
San Diego	0		New York		
San Francisco	39.1	.02	Albany	1907.5	.22
San Pablo	503.1	.02	North Carolina		
Stockton	979.8	.49	Morehead C.	1969.6	.65
Connecticut			Wilmington	3041.6	.39
Bridgeport	224.8	.07	Oregon		
New Haven	566.9	.06	Astoria	881.7	.56
New London	7.8	.00	Coos Bay	3652.3	.66
Delaware[a]	10322.9		Portland	12567.1	.04
Delaware City	516.5		Yaquina Bay	1379.2	19.99
New Castle	1202.2		Pennsylvania		
Wilmington	1851.0	.52	Chester	4.1	
Florida			Marcus Hook	3446.4	
Canaveral	2438.7	.90	Penn Manor	1376.6	
Charlotte	1322.5	.98	Philadelphia	6702.0	.06
Fernandina[b]			Puerto Rico		
Jacksonville	3098.7	.20	Mayaquez	106.7	.28
Key West	25.7	.14	Ponce	77.7	.00
Miami	20.3	.00	San Juan	852.9	.08
Palm Beach	209.9	.13	Rhode Island		
Panama City	210.0	.13	Providence		
Pensacola	633.6	.26	South Carolina		
Port Evergla.	83.7	.00	Charleston	5816.9	
Port St. Joe	167.8	.13	Texas		
Tampa	2309.4	1.90	Beaumont	3990.9	.08
Georgia			Brownsville	3116.6	1.21
Brunswick	3409.5	2.38	Corpus Christi	6202.1	.16
Savannah	10429.8	.85	Freeport	3590.8	.18
Hawaii			Galveston	1638.8	.02
Hilo	352.2	.32	Houston	8312.5	.08
Honolulu	167.7	.03	Port Arthur	3990.9	.13
Kawaihae			Texas City	19545.1	.08
Keweenaw			Virgin Islands		
Nawiliwili	535.3	.68	St. Thomas		
Port Allen	63.7	.62	Virginia		
Louisiana			Newport News	932.0	.04
Baton Rouge	18297.5	.23	Norfolk	2801.7	.05
New Orleans	16661.9	.09	Washington		
Maine			Anacortes		
Portland	613.4	.05	Bellingham	140.8	.08
Searsport			Everett	457.0	.20
Maryland			Grays Harbor	4668.4	1.44
Baltimore	2477.6	.05	Kalama	199.6	.15
Massachusetts			Longview	4645.5	.54
Boston	181.1	.00	Port Angeles	19.3	.00
Fall River	133.2	.03	Seattle	376.5	.02
New Bedford/			Tacoma	44.5	.00
Fairhaven			Vancouver	1185.5	.43
Salem					

[a]See consolidated Delaware River summary.
[b]Maintained by U.S. Navy.

TABLE 19 Federally Funded Maintenance Dredging Projects

Quantity of Materials Dredged (by type of Dredge Employed) (10^3 yds^3)

Area	Bucket	Hopper	Pipeline	All Types Dredged	Pipeline and Bucket	Pipeline and Hopper	Hopper and Bucket	Sub Total
Gulf Coast	3,198	12,550	116,927	0	2,250	0	0	134,917
Pacific Coast	495	2,955	12,497	4,870	0	0	3,160	23,977
Atlantic Coast	5,181	7,258	28,172	3,060	1,375	6,958	4,250	56,255
Great Lakes	481	758	0	1,232	363	675	1,534	5,047
Interior Wtwys	225	0	3,752	0	2,040	0	0	6,017
Sub Total	9,573 (4.2%)	23,522 (10.4%)	161,348 (71.3%)	9,162 (4.1%)	6,029 (2.7%)	7,633 (3.4%)	8,944 (4.0%)	226,214

Number of Jobs

Area	Bucket	Hopper	Pipeline	All Types Dredged	Pipeline and Bucket	Pipeline and Hopper	Hopper and Bucket	Sub Total
Gulf Coast	2	7	74	0	1	0	0	84
Pacific Coast	4	6	18	1	0	0	4	33
Atlantic Coast	11	14	146	3	2	6	5	187
Great Lakes	7	2	0	7	10	5	10	41
Interior Wtwys	1	0	3	0	3	0	0	7
Sub Total	25 (7.1%)	29 (8.2%)	241 (68.5%)	11 (3.1%)	16 (4.5%)	11 (3.1%)	19 (5.4%)	352

SOURCE: U.S. Army Corps of Engineers

211

TABLE 20 Concentrations of Selected Constituents in Dredged Sediments
and Average Global Crustal Materials

Constituent	Dredged Materials Range in Moles Kg^{-1} of Sediment (except as noted)	Average Crustal Materials Range in Moles Kg^1
Trace Metals		
Iron	0.02 – 0.90	0.61 – 1.03
Manganese	$(0.4 – 10) \times 10^{-3}$	$(12 – 18) \times 10^{-3}$
Zinc	$(0.5 – 8) \times 10^{-3}$	$(0.92 – 1.26) \times 10^{-3}$
Copper	$(0.8 – 9400) \times 10^{-6}$	$(460 – 1090) \times 10^{-6}$
Nickel	$(0.2 – 2.6) \times 10^{-3}$	$(0.62 – 1.69) \times 10^{-3}$
Chromium	$(0.02 – 3.8) \times 10^{-3}$	$(0.92 – 1.92) \times 10^{-3}$
Lead	$(5 – 1900) \times 10^{-6}$	$(48 – 77) \times 10^{-6}$
Cadmium	$(0.4 – 600) \times 10^{-6}$	$(0.89 – 1.6) \times 10^{-6}$
Mercury	$(1 – 10) \times 10^{-6}$	$(0.149 – 0.398) \times 10^{-6}$
Synthetic Organic Substances		
Chlorinated Pesticides		0 – 10 mg kg^{-1}
Polychlorinated Biphenyl Compounds		0 – 10 mg kg^{-1}
Other Properties		
pH		6 – 9
Chemical Oxygen Demand		0.03 – 0.04
Oil and Grease		0.1 – 5 g kg^{-1}

SOURCES: Dredged materials adapted from Engler, 1981; average crustal
materials adapted from Rahn, 1976.

TABLE 21   Summary of Biological Effects of Contaminants in Marine Systems

Substance	Water Quality Criteria (mg kg^{-1})	Natural Seawater Concentration (mg kg^{-1})	Most Sensitive Response		Maximum Bioaccumulation	
			Concentration (mg kg^{-1})	Response	Level Reached (mg kg^{-1})	Organism
Cadmium	5000	8	15,000	Retarded sexual development in oysters	1,200	Abalone
Mercury	100	4	5,600	48-h LC_{50} for oyster embryo	7,400	Algae
Copper	0.1x96-h LC_{50}	60	1,000	Reduced phytoplankton growth rate	15,000	Squid
Chromium	0.01x96-h LC_{50}	200	100	Decreased algae growth	260	Zooplankton
DDT	1.0	0	50	Shrimp mortality	10^7 x ambient	Birds
PCBs	1.0	0	300	Sheepshead minnow mortality	10^5 x ambient	Oyster

SOURCE: Kester et al. (1983).

DATE DUE
